FORESTRY COMMISSION BULLETIN
No. 49

The Potential of Western Hemlock, Western Red Cedar, Grand Fir and Noble Fir in Britain

(The 'Minor Species' Project)

By J. R. ALDHOUS, B.A. and A. J. LOW, B.Sc., M.Sc.F., Ph.D.
Forestry Commission

LONDON: HER MAJESTY'S STATIONERY OFFICE
1974

ACKNOWLEDGEMENTS

This report is the result of the work of staff from many sections within the Forestry Commission and of the Princes Risborough Laboratory of the Building Research Establishment, Department of the Environment. Results of surveys carried out by members of specialist sections are appropriately attributed in the text and are warmly acknowledged. However, it is impossible to name all those whose work has been drawn on in the report since this would necessitate listing almost all past and present members of the Silviculture and Genetics research sections. Nevertheless, mention must be made of Messrs. J. Meechan, G. Harrison, J. Jones, A. Little, D. Harrison and A. S. Ford who carried out the crop measurements, site assessments and soil descriptions in the field, Mr. D. F. Fourt who supervised the laboratory handling of soil and assessed the texture of the soils that were analysed, Mrs. M. Cardrick, Miss S. Dabek and Mrs. C. Haggett who undertook the numerous soils analyses, and Mr. D. Durrant who spent much time marshalling and rearranging data and who drew many of the diagrams illustrating the text.

Thanks are also due to those private estate owners, agents and factors who kindly allowed measurements to be taken of their plantations, and also to the Director General of the Meteorological Office for data on rainfall and evapotranspiration for sites where crops were measured.

© *Crown copyright* 1974

ISBN 0 11 710141 9

CONTENTS

			Page
ACKNOWLEDGEMENTS			ii
CHAPTER 1	INTRODUCTION		1
Section:	1.1	Background	1
	1.2	Basis of Report	3
	1.3	Distribution of Minor Species	3
CHAPTER 2	SILVICULTURAL FACTORS IN ESTABLISHMENT AND EARLY GROWTH		5
Section:	2.1	Establishment	5
	2.2	Susceptibility to Late-Spring Frosts	6
	2.3	Early Growth	6
CHAPTER 3	GROWTH AND BEHAVIOUR IN POLE-STAGE AND OLDER CROPS		9
Section:	3.1	Rates of Growth in Pole-Stage and Older Crops	9
	3.2	Stand Structure and Thinning	46
	3.3	Defects Due to Adverse Climatic and Site Factors	47
	3.4	Crop Stability	51
CHAPTER 4	YIELD AND SEED SOURCE: THE NATURAL RANGES OF THE MINOR SPECIES		52
Section:	4.1	Potential for Improvement	52
	4.2	Range of Species in North America	55
CHAPTER 5	DAMAGE BY PESTS		59
Section:	5.1	Fomes Butt Rot	59
	5.2	Honey Fungus	60
	5.3	Other Diseases	60
	5.4	Insect Damage	60
	5.5	Damage by Animals	62
CHAPTER 6	WOOD PROPERTIES, UTILIZATION AND MARKETING		63
Section:	6.1	Wood Properties of Western Hemlock	64
	6.2	Wood Properties of Western Red Cedar	65
	6.3	Wood Properties of Grand Fir	67
	6.4	Wood Properties of Noble Fir	68
	6.5	Timber Prices	70
	6.6	Minimum Quantities for Effective Marketing	70
	6.7	Timber Supplies from Existing Plantations	71
CHAPTER 7	OVERALL COMPARISONS OF SPECIES		73
Section:	7.1	Costs	73
	7.2	Revenues	73
	7.3	Combined Costs and Revenues	74
	7.4	Comparisons of Growth Rates and Valuations	75
CHAPTER 8	FUTURE USE OF SPECIES		78
Section:	8.1	Western Hemlock	78
	8.2	Western Red Cedar	78
	8.3	Grand Fir	79
	8.4	Noble Fir	80
	8.5	Amenity and Recreation	80

Contents *continued*

			Page
SUMMARY			81
REFERENCES			85
APPENDICES			
	I	Crop, Site and Soil Data	87
	II	Report on Survey into Timber Losses in Western Hemlock and Grand Fir as a Result of Infection by *Fomes annosus*, by J. E. Pratt	98
	III	Tree Stability, by J. Everard	101

TEXT FIGURES

			Page
		GRAND FIR COMPARISONS	
(3)	1	Comparison of growth of Sitka spruce and Grand fir	10
,,	2	Comparison of growth of Norway spruce and Grand fir	12
,,	3	Comparison of growth of Douglas fir and Grand fir	14
,,	4	Comparison of growth of Corsican pine and Grand fir	16
		NOBLE FIR COMPARISONS	
,,	5	Comparison of growth of Sitka spruce and Noble fir	18
,,	6	Comparison of growth of Norway spruce and Noble fir	20
,,	7	Comparison of growth of Douglas fir and Noble fir	22
,,	8	Comparison of growth of Corsican pine and Noble fir	24
		WESTERN HEMLOCK COMPARISONS	
,,	9	Comparison of growth of Sitka spruce and Western hemlock	26
,,	10	Comparison of growth of Norway spruce and Western hemlock	28
,,	11	Comparison of growth of Douglas fir and Western hemlock	30
,,	12	Comparison of growth of Corsican pine and Western hemlock	32
		WESTERN RED CEDAR COMPARISONS	
,,	13	Comparison of growth of Sitka spruce and Western red cedar	34
,,	14	Comparison of growth of Norway spruce and Western red cedar	36
,,	15	Comparison of growth of Douglas fir and Western red cedar	38
,,	16	Comparison of growth of Corsican pine and Western red cedar	40
		MISCELLANEOUS COMPARISONS	
,,	17	Comparison of growth of Scots pine and three minor species	42
,,	18	Comparison of growth of Japanese larch and three minor species	44
		NATURAL RANGES	
(4)	19	Natural Range of Sitka spruce, Grand fir, Noble fir	56
,,	20	Natural Range of Western red cedar, Western hemlock and Douglas fir	57
		ADELGID DISTRIBUTION	
(5)	21	Map of the distribution of *Abies grandis* stands infested by *Adelges piceae*	61
(Appendix 3)		RESISTANCE OF TREES TO UPROOTING BY PULLING	
	22	Regression lines of turning moment on stem weight for Grand fir, at all plots studied	103

Contents *continued*

(*Chapter Reference*)	Figure		Page
(Appendix 3)	23	Regression lines of turning moment on stem weight for Western hemlock, at all plots studied	103
	24	Regression lines of turning moment on stem weight for all species studied at Mynydd Du Forest	104
	25	Regression lines of turning moment on stem weight for all species studied at Alice Holt Forest	104

PHOTOGRAPHS

Plate
1. Foliage on lower side branch of Grand fir
2. Fast-growing 33-year-old Grand fir stand in Dunkeld Forest, Perthshire
3. 28-year-old Grand fir stems in Gwydyr Forest, Caernarvonshire
4. 24-year-old Western hemlock stand in Dyfi Forest, Merioneth
5. 50-year-old Western hemlock stems in Langridge Wood, Dunster, Somerset
6. Exposed Noble fir and Grand fir in Beddgelert Forest Garden, Caernarvonshire
7. Drought crack in Noble fir

SELECTED PLANKS SHOWING RANGE OF QUALITY:
8. Grand fir
9. Noble fir
10. Western red cedar
11. Western hemlock

12. Western hemlock and Sitka spruce at Margam Forest, Glamorgan
13. Stem of fluted Western red cedar
14. 28-year-old stand of Western red cedar, Queen Elizabeth Forest, Hampshire
15. 26-year-old stems of Western red cedar, Alice Holt Forest, Hampshire
16. 17-year-old Noble fir plot, Kilmun Forest Garden, Argyll
17. 28-year-old Noble fir stems, Gwydyr Forest, Caernarvonshire
18. Noble fir specimen. Bedgebury Pinetum, Kent.

Chapter 1
INTRODUCTION

1.1 Background

In the second and third decade of this century, the area under forest plantation in Britain was being extended, following Government forest policy and the formation of the Forestry Commission. While the bulk of the afforestation area was planted with Norway and Sitka spruces, Scots and Corsican pines, larches and Douglas fir, a lively interest in the potential of other species led to the planting, in various parts of the country, of a number of trials of "minor" species, among them, Western hemlock, Western red cedar, Grand fir and Noble fir.

At the end of the second world war in 1945, the early growth of these four species appeared sufficiently promising for their planting to be extended, and by the mid-1960s, their growth in some of the earliest plantations was outstanding. Table 1 summarises data from sample plots in some of the most vigorous stands. Many post-war plots were also highly promising. At the same time, however, there were doubts expressed as to the value of the timber from the faster grown plots; defects such as drought crack had frequently been observed on the firs, and the relative susceptibility of different species to *Fomes* heart rot was being investigated.

TABLE 1
GROWTH AND YIELD OF MINOR SPECIES FROM AMONG BEST SAMPLE PLOTS

Forest	Age at last measurement years	Top Ht Main Crop m	Main Crop Volume, over bark cu m/ha	GYC	LYC	Sample Plot No.
				cu m/ha/ann		
WESTERN HEMLOCK						
Laigh of Moray, (Monaughty), Moray and Nairn	43	21·3	441	15	18	3182
Glen Branter, Argyll	32	19·5	305	20	20	3307
Dyfi Corris, Montgomery	44	33·1	566	28	25	2147
Halwill, Devon and Cornwall	34	20·0	644	18	24	1389
WESTERN RED CEDAR						
Gwydyr, Caernarvonshire	39	20·3	452	20	20	2072
Dyfi Corris, Montgomery	40	25·3	543	26	27	2082
Forest of Dean, Gloucestershire	53	25·1	483	20	16	1190
Dunster, Somerset	46	21·3	423	18	14	1391
GRAND FIR						
Laigh of Moray, (Monaughty), Moray and Nairn	43	26·4	535	20	25	3185
Inchnacardoch, (Port Clair), Inverness-shire	36	28·7	633	29	32	3203
Gwydyr, Caernarvonshire	42	32·2	584	27	25	2038
New Forest, Hants.	38	29·4	335	26	19	1449
NOBLE FIR						
Ratagan, Inverness-shire and Ross-shire	36	20·6	533	22	26	3277
Glen Urquhart, Inverness-shire	38	21·0	471	21	21	3306
Gwydyr, Caernarvonshire	42	20·1	440	18	17	2037
Dyfi, Merioneth	45	25·3	461	24	16	2152

Note. GYC, LYC = respectively 'General Yield Class' and 'Local Yield Class' as defined in *Forest Management Tables* (*Metric*) Forestry Commission Booklet No. 34. (HMSO £1.60).

The nomenclature usage in this Bulletin is as follows:

English name	Current Botanical name	Abbreviation in tables
Western hemlock, hemlock:	*Tsuga heterophylla* (Raf.) Sarg.	WH
Western red cedar, Red cedar:	*Thuja plicata* D. Don	RC
Grand fir	*Abies grandis* Lindley	GF
Noble fir	*Abies procera* Rehder	NF
Sitka spruce	*Picea sitchensis* (Bong.) Carr.	SS
Norway spruce	*Picea abies* (L.) Karsten	NS
Corsican pine	*Pinus nigra* var. *maritima* (Aiton) Melville	CP
Scots pine	*Pinus sylvestris* L.	SP
Douglas fir	*Pseudotsuga menziesii* (Mirbel) Franco	DF
Japanese larch	*Larix kaempferi* (Lamb.) Carr.	JL

Consequently, in 1967 it was decided to undertake a comprehensive evaluation of the status and potential of the four most important "minor species" in British forestry, Western hemlock, Western red cedar, Grand fir and Noble fir.

It was immediately apparent that strictly comparable data for at least some major species, i.e. Sitka spruce, Norway spruce, Corsican pine, Scots pine, Douglas fir and Japanese Larch would be essential if a fair evaluation was to be made.

TABLE 2A

AREA OF MINOR SPECIES IN BRITAIN:
FORESTRY COMMISSION PLANTATIONS ESTABLISHED BEFORE 1950
(Source: Working Plan Survey Data)

Hectares

Decade in which trees planted	Conservancy											Great Britain
	England					Scotland				Wales		
	NW	NE	E	SE	SW	N	E	S	W	N	S	
					Western Hemlock							
1930 and older	3·6	0·4	0·8	0	1·2	7·7	7·3	0·8	3·2	13·4	1·6	40
1931–40	3·2	6·5	1·6	0	13·8	56·7	19·4	15·0	37·2	31·2	20·2	204·8
1941–50	16·2	34·8	157·0	39·3	45·0	23·9	2·0	2·8	44·1	32·8	40·9	438·8
Total	23·0	41·7	159·4	39·3	60·0	88·3	28·7	18·6	84·5	77·4	62·7	683·6
	ENGLAND 323 ha					SCOTLAND 220 ha				WALES 140 ha		
					Red Cedar							
1930 and older	0·4	1·6	0·8	1·2	33·2	13·0	0·8	0	12·1	0·4	5·3	68·8
1931–40	1·6	0·4	1·6	1·2	3·2	1·2	2·0	0	4·9	20·2	4·9	41·2
1941–50	0	8·9	2·4	2·4	2·4	2·4	0	0	6·1	4·9	8·9	38·4
Total	2·0	10·9	4·8	4·8	38·8	16·6	2·8	0	23·1	25·5	19·1	148·4
	ENGLAND 61 ha					SCOTLAND 42 ha				WALES 44 ha		
					Grand Fir							
1930 and older	0	0·4	0	0	9·3	4·5	8·1	4·9	1·2	12·6	0	41·0
1931–40	4·5	8·5	3·6	4·9	15·0	15·8	29·5	27·1	8·1	32·4	38·5	187·9
1941–50	1·6	4·5	6·1	8·5	8·5	3·6	0·4	13·0	8·1	3·2	16·2	73·7
Total	6·1	13·4	9·7	13·4	32·8	23·9	38·0	45·0	17·4	48·1	54·7	302·6
	ENGLAND 75 ha					SCOTLAND 124 ha				WALES 103 ha		
					Noble Fir							
1930 and older	0	3·6	0	0	0	9·3	0·8	1·2	13·0	6·9	0	34·8
1931–40	2·4	5·3	0·4	0	0	14·2	1·2	5·3	36·4	2·0	0	67·2
1941–50	0	0	0	0·4	0	0	0	0	6·9	2·8	2·8	12·9
Total	2·4	8·9	0·4	0·4	0	23·5	2·0	6·5	56·3	11·7	2·8	114·9
	ENGLAND 12 ha					SCOTLAND 88 ha				WALES 14 ha		

TABLE 2B

AREA OF MINOR SPECIES IN BRITAIN:
PRIVATELY OWNED PLANTATIONS ESTABLISHED BEFORE 1950

Hectares

	England	Scotland	Wales	Great Britain Total
Western Hemlock	50	38	Nil	88
Red Cedar	129	29	Nil	158
Grand fir	84	30	8	122
Noble fir	40	68	2	110

Work was therefore put in hand to determine:
(i) The age and location of existing stands of the four minor species and of the nearby most productive major species, both in the Forestry Commission and privately owned woodlands.
(ii) The relative performance of each species in terms of establishment costs, survival, rates of growth, quality of timber.
(iii) Whether there were any simple relationships between rates of growth of pairs of species or between species and site characteristics.
(iv) The likely future risks of loss through heart rot, drought crack, windblow and insect attack.
(v) Estimates of differences between past and future performance if present knowledge and capability were applied to plantations to be made in the next twenty years.

The results of these comparisons have, where possible, been expressed in terms of discounted expenditure or discounted revenue.

No allowance has been made in this evaluation for any restraints imposed by amenity considerations, nor for regional differences in timber prices nor such local factors as the disposal of Christmas trees and small dimension produce from forests in close proximity to urban markets.

1.2 Basis of Report

This report is based on the results of detailed assessments in stands selected to represent different sites and growth rates over the country as a whole, as far as this was possible from the limited range of stands of some minor species. These data have been combined with results from over 150 silvicultural experiments. Information on timber properties has also been summarised, using reports from Princes Risborough Laboratory (formerly Forest Products Research Laboratory) as an important source.

Fuller details of sources of information are given in Appendix I, p.87.

1.3 Distribution of Existing Stands in Britain

Table 2a shows the extent of Forestry Commission plantations of the "minor species" dating from 1950 or before. Similar but less detailed data from private woodlands are given in Table 2b. Both tables illus-

TABLE 3
ANALYSIS OF SITES BY SOIL TYPES
NUMBER OF PLOTS OF EACH MINOR SPECIES BY SOIL TYPE

Soil Types	SPECIES				SOIL TYPES INCLUDED UNDER GENERAL HEADINGS
	GF	NF	WH	RC	
Brown Earths	47	24	50	31	Lessivated, Calcareous, Podsolised, Oligotrophic, Eutrophic, Ochreous (plus superficial gleying), Indurated, Imperfectly drained.
Ironpan Soils	3	0	7	1	Iron Podsol, Humus Iron Podsol. Iron Pan with Humus "B"
Surface Water Gley	16	2	13	15	Mineral Gley, Brown Gley, Seepage (flush) Gley (1 SS) Seepage Gley (1 WH) Podsolic Gley (2 WH and 1 RC) Ground Water Gley (1 WH)
Peaty Gley	1	0	1	0	Humic Gley (1 WH and 1 CP)
Deep Peat	0	0	1	1	

Note. Soil types as described by Pyatt (1970).

TABLE 4
PROPORTIONS OF STANDS BY YIELD CLASSES COMPARING SITKA SPRUCE STANDS VISITED IN THE MINOR SPECIES SURVEY WITH THOSE IN THE FORESTRY COMMISSION AS A WHOLE

	Metric Yield Class Sitka spruce											
	28	26	24	22	20	18	16	14	12	10	8	6
	% in each yield class											
Stands selected for comparison with minor species	1	—	2	10	10	19	26	18	10	3	2	
Stands in Forestry Commission as a whole	<1	<1	<1	<1	<1	2	8	19	39	19	10	1

trate how small an area of plantations of the minor species had been established prior to 1950. Areas planted since 1950 are given in Tables 20 and 21. While recent planting greatly exceeds the early plantings, the total area for all minor species combined is not much more than 8 000 hectares.

Table 3 lists the main soil types on which pre-1950 stands of the four minor species have been found and illustrates that between two-thirds and three-quarters of all plots occurred on soils classified as Brown Earths, and that most of the remainder were on Surface Water Gleys. Soil types are as described by Pyatt *et al.* (1969).

The absence of plots on peaty and ironpan soils reflects principally the pattern of planting of minor species in the period 1930–1950 rather than their inability to grow on upland soils. Nevertheless, the results of recent species trials on upland sites offer no justification for believing that there is any large area or type of upland site previously unrecognised, where minor species are likely to grow better than might be predicted from the information already available.

When data on growth of the minor species were collected, each minor species plot was matched with the nearest stand of a major species growing on a similar site. If the surveyed plots of the major species are listed by yield class, it becomes obvious that comparisons with minor species have only been possible on the more productive sites. Table 4 shows, for example, that while most of the Sitka spruce stands selected for comparison with minor species are in Metric Yield Class 14 or better, these are in fact representative of less than one-third of Sitka spruce stands in Britain. More particularly, while there are many examples of the superior growth rates of minor species, especially Grand fir, the more remarkable of these are on exceptionally favourable sites which are extremely limited in extent and represent less than one per cent of land currently available for forestry.

Chapter 2
SILVICULTURAL FACTORS IN ESTABLISHMENT AND EARLY GROWTH

Introduction
Most of the crops available for study were established on bare land as part of an afforestation programme. However, because most of the land currently being afforested is too infertile or climatically unsuited to profitable planting with any of the minor species, the opportunity for their more extensive use comes with the regeneration of the more productive sites already under forest.

The little evidence on the performance of minor species as second generation crops is also summarised in this chapter. Nevertheless, much of the data from crops and experiments established as part of an afforestation programme are equally relevant to the use of minor species in regeneration programmes.

2.1 Establishment
The aspects of establishment affected by choice of species include cost of plants, survival and the need to beat up, weeding and on unfavourable sites where some species develop multiple leaders, the need for singling. Other factors such as cost of land purchase, fencing, ploughing, planting are common to all species and have not been taken into account here.

Table 5 summarises those costs of establishment which are affected by choice of species. All costs are discounted back to the year of planting.

2.11 Cost of Plants
The figures in Table 5 are based on the 1968 Forestry Commission price list and an estimate of the proportions of two-year-old and older plants currently being sent out for planting. The somewhat higher cost of plants of the minor species is due partly to higher costs of seed per 1 000 plants, partly to slow rates of growth of seedlings and

TABLE 5
COMPARATIVE SPECIES-DEPENDENT COSTS OF ESTABLISHMENT
(Discounted £ per hectare)

Operation	SPECIES									
	Western hemlock	Red cedar	Grand fir	Noble fir	Sitka spruce	Norway spruce	Douglas fir	Corsican pine	Scots pine	Japanese larch
Cost of plants	£32	£30	£32	£32	£22	£22	£25	£20	£15	£22
Beating up	£17	£20	£20	£17	£5	£10	£7	£12	£5	£7
Weeding*	£12/50	£25/75	£37/100	£50/125	£12/75	£25/100	£12/75	£25/75	£12/62	£12/50
Sum of costs										
— Range	£61/99	£75/125	£89/152	£99/174	£39/102	£57/132	£44/107	£57/107	£32/82	£41/79
— Median**	£80	£100	£120	£137	£70	£94	£75	£82	£57	£60

* Figures shown as (eg) £12/50 indicate a range between the values given.
** The median cost is the middle value of the range.

TABLE 6
LOSSES AT PLANTING: SUMMARY OF FORESTRY COMMISSION CONSERVANCY AND RESEARCH EVIDENCE

Species	Total No. of Plants Used Annually in Period 1963–7 (All Conservancies)	Mean %** Used for Beating-up	Mean % Loss in Experiments in England and Wales comparing two or more species	
			Open ground	Under Shelter
Western hemlock	2·0 million	29	28(*31*)*	4(*4*)
Western red cedar	0·85 ,,	31	35(*10*)	5(*4*)
Grand fir	0·6 ,,	35	34(*5*)	9(*5*)
Noble fir	0·75 ,,	25	28(*7*)	No data
Sitka spruce	38·0 ,,	5	7(*13*)	1(*2*)
Norway spruce	8·0 ,,	15	21(*5*)	2(*2*)
Douglas fir	3·8 ,,	10	14(*10*)	5(*6*)
Corsican pine	5·2 ,,	20	22(*10*)	0(*3*)

Notes.
* Figures in brackets indicate the number of experiments from which data has been taken to get mean % loss.
** Figures for minor species from special return from conservancies; figures for major species estimated.

transplants and partly to a lower survival rate following transplanting.

2.12 Survival and Beating Up

Recent experience on planting losses is summarised in Table 6. The higher beating up costs in Table 5 reflect the fact that for the minor species, planting losses and consequent beating up requirements are often relatively high, averaging between 25 and 35 per cent of plants used, compared with from 5 to 20 per cent for most major species on similar sites.

The difference in survival between trees in open ground and under shelter in experiments in England and Wales is remarkable. Evidence that hemlock responds to shelter is supported by post-war experience of planting into light scrub cover in South-east England.

In Scottish silvicultural experiments, similar results have been obtained though the losses have been generally lower overall. On deep peats and peaty gleys, survival of the minor species has been variable and generally poorer than that of Lodgepole pine, Sitka spruce and larches.

In respect of establishment, Western hemlock has been the most satisfactory of the minor species while Red cedar in Scotland has frequently failed. There is no evidence that Noble fir anywhere has any advantage in survival over Sitka spruce or Lodgepole pine. In high elevation experiments and on exposed sites, it has not normally survived so well.

2.13 Weeding Costs and Weeding

The costs shown in Table 5 are dependent principally on rates of growth in the first two or three years following planting but partly also on predicted weedgrowth. Western hemlock is the exception among the minor species in this respect, being quickly established and soon growing out of the weeds, in particular when raised under partial shade. Noble fir is at the other extreme.

Western hemlock has been found to be more susceptible to 2,4-D than Sitka spruce and less easily freed from *Calluna* competition.

2.2 Susceptibility to Late Spring Frosts

Grand fir and to a lesser extent Western hemlock are relatively susceptible to late spring frosts which leave Scots pine undamaged and only slightly harm Corsican pine. This susceptibility is sufficient to make it essential that Grand fir be protected in some way, if it is to be established where late spring frosts may be expected. Areas such as East Anglia and the East Midlands have such a history of late spring frosts that there is no reasonable prospect of establishing Grand fir unless there is some protection against frost. Table 7 illustrates the contrast between the susceptibility of Corsican pine and Grand fir to late spring frosts in Thetford Forest, and the sort of overcrop that is required to provide protection to trees which had been planted two years previously.

Red cedar also appears to be susceptible to spring frost though markedly less susceptible than Grand fir or Western hemlock. However, because Red cedar has no clearly visible bud or rapidly elongating succulent shoot, damage cannot immediately be distinguished and is only apparent when frost cankers appear later.

Frost damage may be more pronounced in the second year after planting than in the first year, when the check of planting, by delaying bud-break often enables new shoots to avoid damage by late spring frost.

2.21 Overhead Cover as Protection Against Radiation Frosts

Observations in plantations show that overhead shelter from older trees can result in temperatures being between 3 and 5°C higher under cover than on open land. In the late spring, this difference can often be critical for susceptible species. Table 7 illustrates how Grand fir can benefit from overhead shelter on a frosty site.

TABLE 7

FROST DAMAGE TO YOUNG CORSICAN PINE AND GRAND FIR UNDERPLANTED IN 1966 UNDER SCOTS PINE, THETFORD FOREST (FELTWELL BEAT) ASSOCIATED WITH A GROUND FROST OF −6° C IN LATE MAY, 1968

No. of stems per hectare of overcrop Scots pine	0	80	185	392
% of Corsican pine damaged	0	0	0	0
% of Grand fir damaged	48	47	26	0

While at first sight it may appear uneconomic to provide shelter of this sort, the data from silviculture experiments indicate that the value and volume increment of remaining trees is sufficient to cover any financial disadvantage resulting from deferred realisation of these trees.

In the absence of direct experience, it is considered that provided the trees in the overcrop are left in rows for between five and ten years, any additional harvesting costs are likely both to be small and to be balanced by savings due to the reduced amount of brash on the site when the bulk of the trees are felled. Therefore, no firm basis exists for ascribing either a net cost or a net revenue to the provision of overhead cover when establishing minor species.

2.3 Early Growth

2.31 Western hemlock

There are nearly 100 experimental comparisons of the growth of hemlock with one or other major species. It is at once obvious that the rate of growth of hemlock on sheltered lowland sites is superior to

TABLE 8
EARLY GROWTH OF SPECIES ON FAVOURABLE SITES
Sheltered Brown Earths

Mean Height in Metres

Forest	Age yrs.	Western hemlock	Red cedar	Grand fir	Noble fir	Sitka spruce	Norway spruce	Douglas fir	Hybrid larch	Japanese larch
Kernow: (Bodmin, Glynn) Expt. 2	5	1·7	—	0·7	—	1·1	—	—	1·7	—
Tair Onen (Michaelston) " 3	4	2·1	—	0·6	—	1·5	0·9	1·4	—	2·6
Radnor 35	6	2·2	1·2	1·2	—	1·8	1·2	1·6	—	—
Ae 19	6	1·3	0·8	0·4	0·6	1·8	0·5	1·0	—	—
Glen Righ 39	6	1·8	—	0·7	0·9	1·9	0·9	2·1	—	—
Inchnacardoch (Port Clair) 3	6	2·0	—	0·9	0·9	1·6	1·0	2·3	—	—

that of all other conifers except larch, whether the shelter is provided by other trees or by the local topography. On such sites, it is not uncommon for the faster individuals in a hemlock crop to reach 1.8 m in four years and for the whole crop to average 1.8 m in six years. (Table 8).

Douglas fir almost everywhere grows less rapidly than hemlock on similar sites. This relationship is remarkably consistent, the only exception being on calcareous sites, such as shallow sand over chalky boulder clay at Thetford Forest and on shallow loam over chalk at Gardiner Forest, where hemlock grows slowly.

The early growth of hemlock in comparison with Sitka and Norway spruce varies very markedly according to the site, these species having been planted on many more difficult sites than have been planted with Douglas fir. On the best sites, hemlock grows away quickly, a little faster than Sitka spruce on some sites and a little slower than Sitka spruce on others (Table 8) though everywhere hemlock is very much faster than Norway spruce.

Site factors which alter this relationship are exposure, elevation and infertility. On upland sites where spruces are not checked by low fertility, Sitka spruce grows somewhat faster, has a better survival rate, is more resistant to blast and does not fork. In comparison with Norway spruce, hemlock has grown as fast on more elevated sites but the rate of growth of both species has usually been poor, i.e. 15–20 cm per annum over the first ten years.

Hemlock, while surviving and growing on many upland sites is very often cut back by cold winter winds and develops into a bushy plant with many leaders, subsequently growing into pole crops with many more multiple-stemmed trees than are ever found in a Sitka spruce stand (Plate 7).

Where sites are infertile but not unduly high or exposed, hemlock has grown well, whereas spruces have been slow and on a number of sites have gone into check. On such areas, hemlock may be several times taller than the spruce, the contrast being more accentuated, the more seriously the spruce is checked.

Where Corsican pine and hemlock have been planted together, the hemlock gets away much more quickly, often requiring one fewer weeding than the pine.

2.32 Other Minor Species
There are far fewer comparisons of the other minor species than there are of hemlock.

Red cedar is initially slower on almost all sites than Douglas fir and Sitka spruce. On lowland sites, it is a little quicker off the mark than Norway spruce, but is a little slower on intermediate upland sites. On the more exposed upland sites, Red cedar generally fails.

Grand fir and Noble fir are consistently the slowest to get away on all sites except where spruces are in check, Noble fir being behind Grand fir on most lowland sites.

Chapter 3
GROWTH AND BEHAVIOUR IN POLE-STAGE AND OLDER CROPS

Introduction

During this stage, minor species on the better sites are able to grow very rapidly and offset slow early growth. Comparisons of yield class between minor and the most appropriate of the four major species on more fertile sites usually show the minor species to be the higher yielding.

Some defects can become apparent at this stage. Of these, drought crack is the most serious, but poor stem form and crown blast can also cause loss of value locally.

3.1 Rates of Growth in Pole-stage and Older Crops

The comparisons of rates of growth of plots on similar sites are presented graphically in Figs. 1–18. Each of these diagrams shows the yield class of one of four principal major species on the vertical axis and the yield class of the particular minor species on the horizontal axis. A line is drawn at 45°, dividing the figure into two. Points falling in the lower (or right hand) part of the figure refer to pairs of plots where the minor species is growing at a rate indicating a higher yield class than the major species; points falling in the upper (or left hand) part of the figure describe pairs of plots where the major species is the more productive.

The measure of growth used for comparison is the Production Class or Local Yield Class (Hamilton and Christie, 1971) which in the present study has given rise to slightly more consistent relationships than have General Yield Class figures. The conclusions from each species comparison and its supporting data are summarised in Table 9.

TABLE 9
SUMMARY OF RELATIVE GROWTH RATES OF SPECIES

	Sitka spruce (SS)	Norway spruce (NS)	Douglas fir (DF)	Corsican pine (CP)
Grand fir (GF)	Comparisons available only on sites YC 14 and above. At lower end of range growth rates are similar; the better the site, the greater the relative advantage of Grand fir. (16 comparisons – fig. 1)	GF 2–4 Yield Classes better than NS over whole range of available comparisons. (19 comparisons – fig. 2)	As for SS. (19 comparisons – fig. 3)	GF several Yield Classes more productive on most sites, advantage increasing the better the site. (11 comparisons – fig. 4)
Noble fir (NF)	On two-thirds of the total number of sites, NF is 1–3 Yield Classes better than SS, but on some SS is better. No clear trend in relation to site quality. On average, NF produces 1 Yield Class more than SS. (15 comparisons – fig. 5)	NF slightly higher volume on most sites. Where NS exposed, NF much superior. Suggestions that NF is at a greater advantage in Scotland than further south. (9 comparisons – fig. 6)	Differences small. (6 comparisons – fig. 7)	Insufficient data. Differences observed small. (2 comparisons – fig. 8)
Western hemlock (WH)	WH slightly more productive on the majority of sites studied. SS better in exposed conditions, but WH better on sites of low fertility which are not exposed. (18 comparisons – fig. 9)	WH on average 2 Yield Classes better than NS. (20 comparisons – fig. 10)	Diverse relationships. In Scotland, little to choose in yield. (6 comparisons). In England and Wales, WH on average 2 Yield Classes better than DF. (15 comparisons). (21 comparisons in all – fig. 11)	WH slightly better on more productive sites. CP more productive on least fertile low elevation sites. (6 comparisons – fig. 12)
Red cedar (RC)	Less consistent relationship than for other pairs of species. RC better on lowland sites. SS better above 275 m in England and Wales and 150 m in Scotland. (10 comparisons – fig. 13)	No difference in Yield Class. No data from Scotland. (7 comparisons – fig. 14)	Trend for RC to be increasingly more productive than DF the better the site. The 4 comparisons from Scotland show RC with smaller advantage than in England and Wales. (20 comparisons – fig. 15)	On heavy clay soils and surface water gleys, RC 3–4 Yield Classes better than CP. (4 comparisons). On freely drained soils, S and C England, species equal in production. (4 comparisons). No data from Scotland. At one site on Borders CP markedly more productive than RC. (9 comparisons – fig. 16)

Figs. 17 and 18 give what very little data was collected for Scots pine and Japanese larch in comparison with the minor species.

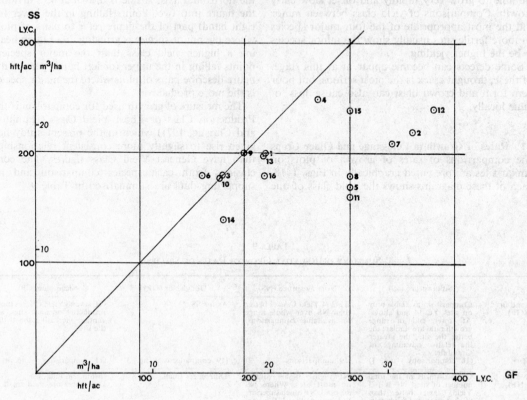

FIGURE 1. COMPARISON OF GROWTH OF SITKA SPRUCE AND GRAND FIR

Each point represents a site where both species are growing. Points shown above the line at 45° across the diagram represent those sites where Sitka spruce is growing faster than Grand fir; conversely, for points below the line, Grand fir is the faster species.

Regression data: metric yield classes
$(y = a + bx)$

y	x	a	b	Correlation Coefficient
SS LYC	GF LYC	10·8	0·34	0‡64**
‡GF LYC	SS LYC	0·80	1·19	

‡ figure reversed.

Key to fig. 1 **Location and Detailis of Plots Comparing Growth Rates of Sitka Spruce and Grand Fir**

Number on Diagram	Species	Forest Name	Compt. No.	Planting Year	Local Yield Class	Soil Type	Survey Stand No.
1	GF	Halwill,	62/4/4d	1938	20	Surface Water Gley	51
	SS	Devon	62/4/4e	1938	18	,, ,, ,,	49
2	GF	Quantock,	49/3/23b	1927	32	Oligotrophic Brown Earth	66
	SS	Somerset	49/3/22	1924	20	,, ,, ,,	67
3	GF	Crychan,	25a	1937	16	Ochreous Brown Earth*	83
	SS	Carmarthen	25b	1935	16	Ochreous Brown Earth	84
4	GF	Tair Onen,	11f	1941	24	Oligotrophic Brown Earth	91
	SS	Glamorgan	10f	1925	22	,, ,, ,,	92
5	GF	Cynwyd,	(54)	1928	26	Ochreous Brown Earth	132
	SS	Merioneth	(54)	1928	16	Surface Water Gley	131
6	GF	Radnor,	56d	1939	14	Ochreous Brown Earth*	165
	SS	Radnor	58b	1939	16	,, ,, ,,	166
7	GF	Dean (Blakeney),	444g	1929	30	Oligotrophic Brown Earth	179
	SS	Glos.	444h	1918	18	,, ,, ,,	180
8	GF	Dean (Churchill),	246g	1938	26	Brown Gley	185
	SS	Glos.	246d	1936	16	,, ,,	187
9	GF	Bedgebury,	Plot No 28	1931	18	Brown Gley	252
	SS	Kent	Plot No 69	1949	18	,, ,,	255
10	GF	Ratagan,	37i	1932	16	Indurated Brown Earth	504
	SS	Ross.	37j	1932	16	Brown Earth	505
11	GF	Glen Urquhart,	25f	1935	26	Ochreous Brown Earth	512
	SS	Inverness	25a/g	1935	14	Brown Earth*	513
12	GF	Inchnacardoch (Port	19d	1932	34	Brown Earth	533
	SS	Clair), Inverness	24b	1929	22	,, ,,	534
13	GF	Inchnacardoch(Port	71d	1933	20	Brown Earth	535
	SS	Clair), Inverness	71b	1933	18	,, ,,	536
14	GF	Montreathmont	19	1938	16	Imperfectly drained	637
	SS	(Inglismaldie), Angus	18	1938	12	Brown Earth	636
15	GF	Laigh of Moray	72	1928	26	Lessivated Brown Earth	641
	SS	(Monaughty), Moray	81	1929	22	Brown Earth*	639
16	GF	Glenfinart,	20	1941	20	Ochreous Brown Earth*	720
	SS	Argyll	16	1941	16	,, ,, ,,	721

*With superficial gleying.
Plots are listed in numerical order in Appendix I, page 87.

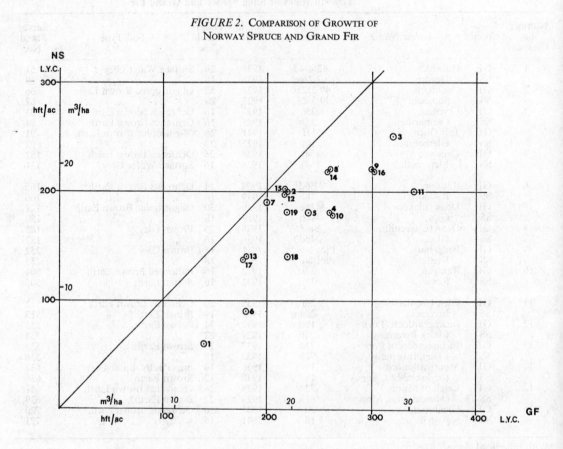

FIGURE 2. COMPARISON OF GROWTH OF NORWAY SPRUCE AND GRAND FIR

Each point represents a site where both species are growing. Points shown above the line at 45° across the diagram represent those sites where Norway spruce is growing faster than Grand fir; conversely, for points below the line, Grand fir is the faster species.

Regression data: metric yield classes
$(y = a + b x)$

y	x	a	b	Correlation Coefficient
NS LYC	GF LYC	0·37	0·76	0·76***
‡GF LYC	NS LYC	8·79	0·75	

‡ figure reversed.

Key to fig. 2 **Location and Details of Plots Comparing Growth Rates of Norway Spruce and Grand Fir**

Number on Diagram	Species	Forest Name	Compt. No.	Planting Year	Local Yield Class	Soil Type	Survey Stand No.
1	GF	Chiddingfold,	10	1939	12	(Non Peaty	11
	NS	Surrey and Sussex	10b	1934	6	(Surface Water Gley	12
2	GF	Chiddingfold,	68c	1943	20	(Non Peaty	15
	NS	Surrey and Sussex	69c	1947	18	(Surface Water Gley	17
3	GF	Mynydd Du,	22d	1941	28	(Oligotrophic	100
	NS	Brecon and Mon.	31a	1944	22	(Brown Earth	102
4	GF	Talybont,	3c	1942	24	(Oligotrophic	108
	NS	Brecon	3d	1948	16	(Brown Earth	110
5	GF	Talybont,	4c	1942	22	(Oligotrophic	109
	NS	Brecon	3d	1948	16	(Brown Earth	110
6	GF	Bagley Wood,	25e	1904	16	(Non Peaty	176
	NS	Oxford	16	1907	8	(Surface Water Gley	175
7	GF	Bedgebury,	28	1931	18	Brown Gley	252
	NS	Kent	15	1932	18	,, ,,	254
8	GF	Inchnacardoch,	36a	1928	24	Brown Earth	520
	NS	Inverness	36e	1932	20	,, ,,	521
9	GF	Ratagan,	40bk	1932	26	Brown Earth	537
	NS	Inverness and Ross	40bk	1932	20	,, ,,	538
10	GF	Inchnacardoch,	49d	1934	24	Brown Earth	544
	NS	Inverness	49e	1922	16	,, ,,	545
11	GF	Dunkeld (Craigvinean),	83	1940	20	Brown Earth	614
	NS	Perth	83	1940	18	,, ,,	615
12	GF	Drummond Hill,	97	1937	20	Brown Earth	624
	NS	Perth	57	1937	18	,, ,,	625
13	GF	Mearns (Drumtochty), Kincardine	66b	1929	16	Brown Earth with Superficial Gleying	630
	NS		66d	1929	12	Humus Iron Podzol	631
14	GF	Achaglachgach,	41/44	1939	24	Brown Earth	707
	NS	Argyll	41/44	1939	20	,, ,,	708
15	GF	Eredine, Argyll	80	1939	20	Brown Earth with Superficial Gleying	713
	NS		80	1939	20	Brown Earth with Superficial Gleying	714
16	GF	Solway (Mabie),	24	1949	26	Brown Earth	820
	NS	Kirkcudbright & Dumfries	24	1949	20	,, ,,	821
17	GF	Allerston District,	SS25e	—	16	Humus Iron Podzol	903
	NS	Yorkshire	—?	1946	12	,, ,, ,,	902
18	GF	Kielder District,	1330	1939	20	Peaty Gley	916
	NS	Northumberland	1330	1939	12	,, ,,	917
19	GF	Kielder District,	1144	1937	20	Mineral Gley	918
	NS	Northumberland	1144	1937	16	,, ,,	919

Plots are listed in numerical order in Appendix I.

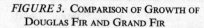

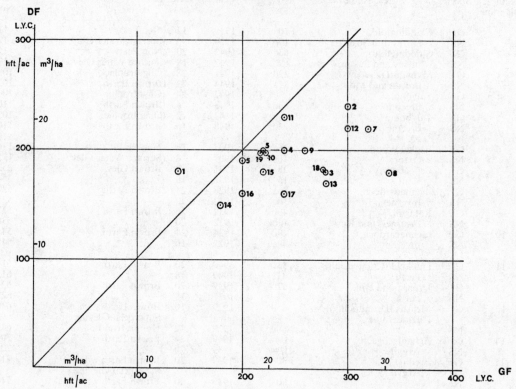

Each point represents a site where both species are growing. Points shown above the line at 45° across the diagram represent those sites where Douglas fir is growing faster than Grand fir; conversely, for points below the line, Grand fir is the faster species.

Regression data: metric yield classes
$(y = a + bx)$

y	x	a	b	Correlation Coefficient
DF LYC	GF LYC	12·9	0·19	0·40
GF LYC	DF LYC	7·83	0·83	

‡ figure reversed.

Key to fig. 3 **Location and Details of Plots Comparing Growth Rates of Douglas Fir and Grand Fir**

Number on Diagram	Species	Forest Name	Compt No.	Planting Year	Local Yield Class	Soil Type	Survey Stand No.
1	GF	Bedgebury (Vinehall),	23	1950	12	(Non Peaty	30
	DF	Sussex	29	1952	16	(Surface Water Gley	31
2	GF	Dartington Estate,	1	1934	26	Eutrophic Brown Earth	37
	DF	Devon	IV	1942	22	,, ,, ,,	39
3	GF	Mendip,	194a	1933	24	(Non Peaty	52
	DF	Somerset & Wilts	194c	1934	16	(Surface Water Gley	53
4	GF	Dartmoor (Plym),	15	1951	22	Ochreous Brown Earth	60
	DF	Devon	7	1954	18	,, ,, ,,	62
5	GF	Dartmoor (Plym),	8	1950	20	,, ,, ,,	63
	DF	Devon	7	1954	18	,, ,, ,,	62
6	GF	Brechfa,	833d	1938	18	(,, ,, ,,	75
	DF	Carmarthen	833c	1938	18	(with Superficial Gleying	76
7	GF	Mynydd Du,	22d	1941	28	Oligotrophic Brown Earth	100
	DF	Brecon & Mon.	22a	1940	20	,, ,, ,,	99
8	GF	Coed-y-Brenin,	R15d	1931	30	Ochreous Brown Earth	125
	DF	Merioneth	R15a	1927	16	,, ,, ,,	127
9	GF	Coed-y-Brenin,	P7f	1940	24	Ochreous Brown Earth	128
	DF	Merioneth	P8a	1936	18	,, ,, ,,	129
10	GF	Dyfi (Corris),	41	1928	20	,, ,, ,,	138
	DF	Merioneth & Mont.	41b	1928	18	,, ,, ,,	140
11	GF	Gwydyr,	199b	1927	22	Podzolised Brown Earth	151
	DF	Caerns. & Denbs.	199b	1927	20	,, ,, ,,	152
12	GF	Gwydyr,	206e	1930	26	Eutrophic Brown Earth	153
	DF	Caerns. & Denbs.	206b	1930	20	,, ,, ,,	154
13	GF	Gwydyr,	320	1927	26	Ochreous Brown Earth	155
	DF	Caerns. & Denbs.	321h	1921	16	,, ,, ,,	158
14	GF	Bagley Wood,	25e	1904	16	(Non Peaty Surface	176
	DF	Oxfordshire	16	1907	14	(Water Gley	174
15	GF	Dean,	80c	1932	20	Podzolised Brown Earth	195
	DF	Glos, Hereford & Mon.	80b	1927	16	,, ,, ,,	196
16	GF	Bedgebury,	28	1931	18	Brown Gley	252
	DF	Kent/Sussex	98	1931	14	,, ,,	256
17	GF	Wensum,	36	1934	22	Humus Iron Podzol	230
	DF	Norfolk	31	1934	14	,, ,, ,,	231
18	GF	Inchnacardoch,	14g	1934	26	Brown Earth	531
	DF	(Portclair), Inverness	14b	1927	16	,, ,,	529
19	GF	Alltcailleach,	22	1932	18	Brown Earth	601
	DF	Aberdeen	22	1932	18	,, ,,	602

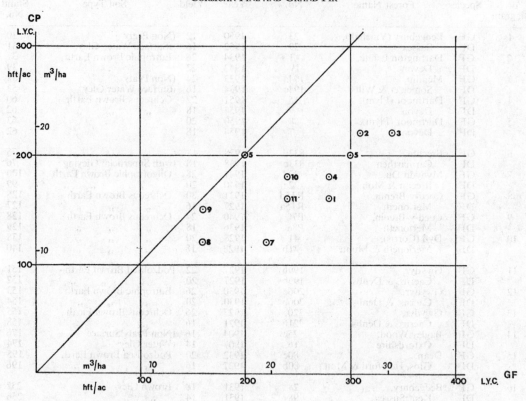

FIGURE 4. COMPARISON OF GROWTH OF CORSICAN PINE AND GRAND FIR

Each point represents a site where both species are growing. Points shown above the line at 45° across the diagram represent those sites where Corsican pine is growing faster than Grand fir; conversely, for points below the line, Grand fir is the faster species.

Regression data: metric yield classes
$(y = a + b x)$

y	x	a	b	Correlation Coefficient
CP LYC	GF LYC	5·96	0·43	0·73***
‡GF LYC	CP LYC	3·02	1·23	

‡ figure reversed.

Key to fig. 4 **Locations and Details of Plots Comparing Growth Rates of Corsican Pine and Grand Fir**

Number on Diagram	Species	Forest Name	Compt. No.	Planting Year	Local Yield Class	Soil Type	Survey Stand No.
1	GF	Alice Holt	63c	1941	24	(Non Peaty Surface	1
	CP	Hampshire/Surrey	62e	1942	14	(Water Gley	2
2	GF	St. Leonards	329(i)	1948	28	Surface Water Gley	20
	CP	(Maresfield), Sussex	336(a)	1942	20	,, ,, ,,	21
3	GF	Ringwood,	29f	1940	30	Podzolised Brown Earth	28
	CP	Hants	28	1939	20	,, ,, ,,	29
4	GF	Gwydyr,	320	1927	26	Ochreous Brown Earth	155
	CP	Caerns. & Denbs.	321	1931	16	,, ,, ,,	159
5	GF	Dean,	246g	1938	26	Brown Gley	185
	DF	Glos., Hereford & Mon.	245c	1934	18	Oligotrophic Brown Gley	188
6	GF	Bedgebury,	28	1931	18	Brown Gley	252
	CP	Kent/Sussex	89	1934	18	,, ,,	257
7	GF	Ampthill,	2	1950	20	(Non Peaty Surface Water	222
	CP	Beds & Herts	5	1951	8	(Gley	223
8	GF	Thetford,	55	1932	14	Iron Podzol	225
	CP	Norfolk/Suffolk	16	1927	10	Podzolised Brown Earth	224
9	GF	Thetford,	157	1951	14	Podzolised Brown Earth	226
	CP	Norfolk/Suffolk	158	1927	14	,, ,, ,,	229
10	GF	Wensum,	36	1934	22	Humus Iron Podzol	230
	CP	Norfolk	33	1934	16	Surface Water Gley	232
11	GF	Wensum,	26	1944	22	Oligotrophic Brown Earth	233
	CP	Norfolk	26	1944	14	,, ,, ,,	234

FIGURE 5. COMPARISON OF GROWTH OF
SITKA SPRUCE AND NOBLE FIR

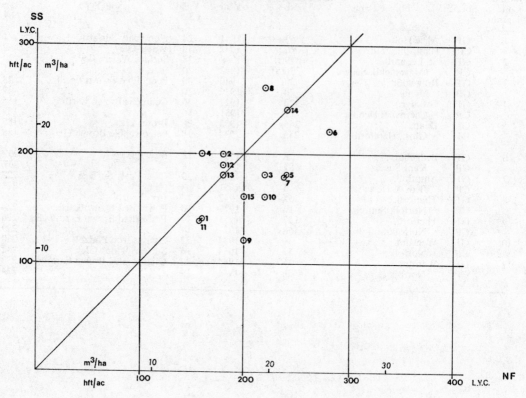

Each point represents a site where both species are growing. Points shown above the line at 45° across the diagram represent those sites where Sitka spruce is growing faster than Noble fir; conversely, for points below the line, Noble fir is the faster species.

Regression data: metric yield classes
$(y = a + bx)$

y		x		a	b	Correlation Coefficient
SS	LYC	NF	LYC	7·85	0·46	0·44
‡NF	LYC	SS	LYC	11·5	0·42	

‡ figure reversed.

Key to fig. 5 **Location and Details of Plots Comparing Growth Rates of Sitka Spruce and Noble Fir**

Number on Diagram	Species	Forest Name	Compt. No.	Planting Year	Local Yield Class	Soil Type	Survey Stand No.
1	NF	Dyfi (Corris),	95c	1935	14	(Ochreous Brown Earth	141
	SS	Merioneth. & Mont.	95b	1935	12	(with Superficial Gleying	142
2	NF	Dyfi (Valley),	13a	1927	18	Ochreous Brown Earth with Superficial Gleying	147
	SS	Merioneth. & Mont.	13b	1927	18	Ochreous Brown Earth	148
3	NF	Radnor,	73c	1928	20	(Ochreous Brown Earth	164
	SS	Rads. & Hereford	72b	1928	14	(with Superficial Gleying	163
4	NF	Bedgebury,	38	1931	14	Brown Gley	253
	SS	Kent & Sussex	69	1949	18	,, ,,	255
5	NF	Glen Urquhart	5f	1931	22	Brown Earth	518
	SS	Inverness	5b	1931	16	,, ,,	519
6	NF	Ratagan,	40ac	1932	26	(Brown Earth with	540
	SS	Inverness & Ross	43g	1931	20	(Superficial Gleying	541
7	NF	Glenbranter,	48	1928	22	Ochreous Brown Earth	718
	SS	Argyll	48	1928	16	,, ,, ,,	719
8	NF	Knapdale,	13d	1943	20	Brown Earth	727
	SS	Argyll	13a	1943	24	,, ,,	726
9	NF	Glenbranter,	96	1935	18	Ochreous Brown Earth	739
	SS	Argyll	96	1935	10	Humus Iron Podzol	740
10	NF	Glenbranter,	7	1937	20	Brown Earth	741
	SS	Argyll	7	1937	14	Podzolised Brown Earth	742
11	NF	Fearnoch,	27	1935	14	Brown Earth	746
	SS	Argyll	27	1935	12	,, ,,	746
12	NF	Gwydyr,	322a	1924	18	Ochreous Brown Earth	150
	SS	Caerns. & Denbs.	320	1927	16	,, ,, ,,	156
13	NF	Dean (Churchill),	254c	1933	16	Brown Gley	182
	SS	Glos., Hereford & Mon.	246d	1938	16	,, ,,	187
14	NF	Inchnacardoch	14f	1934	22	Brown Earth	532
	SS	(Portclair), Inverness	24b	1929	24	,, ,,	534
15	NF	Benmore,	133	1916	18	Ochreous Brown Earth	709
	SS	Argyll	133	1916	14	,, ,, ,,	710

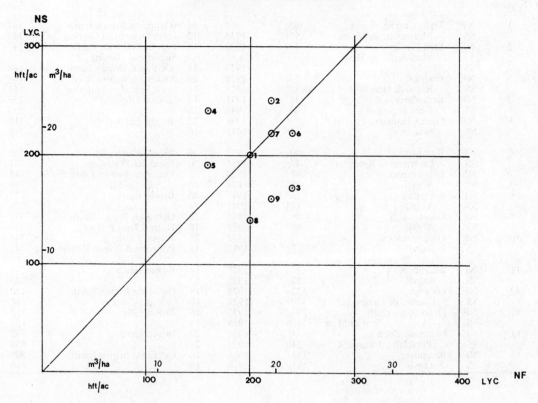

FIGURE 6. COMPARISON OF GROWTH OF
NORWAY SPRUCE AND NOBLE FIR

Each point represents a site where both species are growing. Points shown above the line at 45° across the diagram represent those sites where Norway spruce is growing faster than Noble fir; conversely for points below the line, Noble fir is the faster species.

Regression data: metric yield classes
$(y = a + bx)$

y	x	a	b	Correlation Coefficient
NS LYC	NF LYC	20·1	−0·13	−0·10
‡NF LYC	NS LYC	19·9	−0·48	

‡ figure reversed.

Key to fig. 6 **Location and Details of Plots Comparing Growth Rates of Norway Spruce and Noble Fir**

Number on Diagram	Species	Forest Name	Compt. No.	Planting Year	Local Yield Class	Soil Type	Survey Stand No.
1	NF	Tair Onen,	18d	1941	18	Podzolised Brown Earth	93
	NS	Glamorgan	18d	1941	18	,, ,, ,,	94
2	NF	Mynydd Du	22b	1941	20	Oligotrophic Brown Earth	101
	NS	Brecon & Mon.	31a	1944	22	,, ,, ,,	102
3	NF	Talybont,	37b	1943	22	Oligotrophic Brown Earth	114
	NS	Brecon	37a	1943	16	,, ,, ,,	115
4	NF	Dyfi (Corris),	95c	1935	14	(Ochreous Brown Earth	141
	NS	Merioneth. & Mont.	30a	1933	22	(with Superficial Gleying	144
5	NF	Bedgebury,	38	1931	14	Brown Gley	253
	NS	Kent & Sussex	15	1932	18	,, ,,	254
6	NF	Inchnacardoch, Inverness	36f	1932	22	Imperfectly Drained Brown Earth	522
	NS		36e	1932	20	Brown Earth	521
7	NF	Ratagan, Inverness & Ross	40 bj	1932	20	Indurated Brown Earth with Superficial Gleying	539
	NS		40 bk	1932	20	Brown Earth	538
8	NF	Mearns (Drumtochty), Kincardine	66c	1929	18	Brown Earth with Superficial Gleying	632
	NS		66d	1929	12	Humus Iron Podzol	631
9	NF	Queensberry Eastate	—	1939	20	Brown Earth	824
	NS		—	1939	14	,, ,,	823

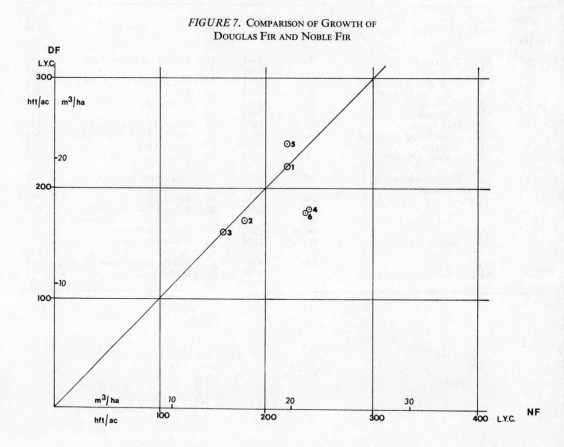

FIGURE 7. COMPARISON OF GROWTH OF DOUGLAS FIR AND NOBLE FIR

Each point represents a site where both species are growing. Points shown above the line at 45° across the diagram represent those sites where Douglas fir is growing faster than Noble fir; conversely, for points below the line, Noble fir is the faster species.

Regression data: metric yield classes
$(y = a + b x)$

y	x	a	b	Correlation Coefficient
DF LYC	NF LYC	9·11	0·43	0·49
‡NF LYC	DF LYC	10·7	0·47	

‡ figure reversed.

Key to fig. 7 **Location and Details of Plots Comparing Growth Rates of Douglas Fir and Noble Fir**

Number on Diagram	Species	Forest Name	Compt. No.	Planting Year	Local Yield Class	Soil Type	Survey Stand No.
1	NF	Mynydd Du,	22b	1941	20	Oligotrophic Brown Earth	101
	DF	Brecon & Mon.	22a	1940	20	,, ,, ,,	99
2	NF	Gwydyr,	320	1927	16	Ochreous Brown Earth	156
	DF	Caerns. & Denbs.	321h	1921	16	,, ,, ,,	158
3	NF	Bedgebury	38	1931	14	Brown Gley	253
	DF	Kent & Sussex	98	1931	14	,, ,,	256
4	NF	Glen Urquhart,	69bc	1929	22	Ochreous Brown Earth	514
	DF	Inverness	65fg	1928	16	,, ,, ,,	515
5	NF	Glen Urquhart,	5f	1931	20	Brown Earth	516
	DF	Inverness	5h	1931	22	,, ,,	517
6	NF	Inchnacardoch	14f	1934	22	Brown Earth	532
	DF	(Portclair), Inverness	14b	1934	16	,, ,,	529

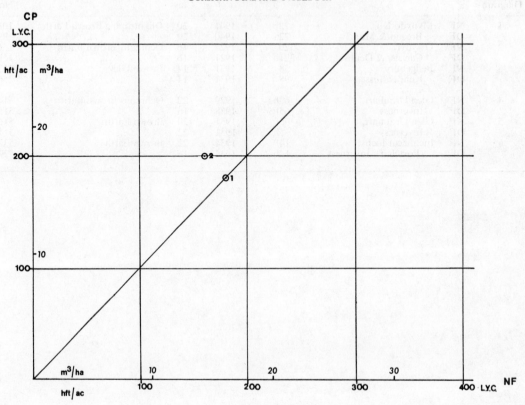

FIGURE 8. COMPARISON OF GROWTH OF CORSICAN PINE AND NOBLE FIR

Each point represents a site where both species are growing. Points shown above the line at 45° across the diagram represent those sites where Corsican pine is growing faster than Noble fir; conversely, for points below the line, Noble fir is the faster species.

Regression data: metric yield classes
$(y = a + b x)$

y	x	a	b	Correlation Coefficient
CP LYC	NF LYC	Insufficient data		Insufficient data
NF LYC	CP LYC			

Key to fig. 8 **Location and Details of Plots Comparing Growth Rates of Corsican Pine and Noble Fir**

Number on Diagram	Species	Forest Name	Compt. No.	Planting Year	Local Yield Class	Soil Type	Survey Stand No.
1	NF	Gwydyr,	320	1927	16	Ochreous Brown Earth	156
	CP	Caerns. & Denbs.	321	1931	16	,, ,, ,,	159
2	NF	Bedgebury	38	1931	14	Brown Gley	253
	CP	Kent & Sussex	89	1934	18	,, ,,	257

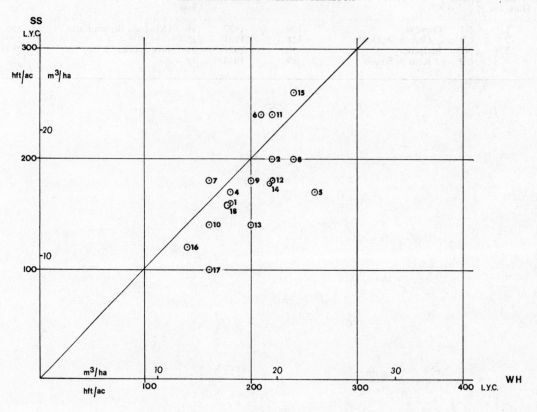

FIGURE 9. COMPARISON OF GROWTH OF SITKA SPRUCE AND WESTERN HEMLOCK

Each point represents a site where both species are growing. Points shown above the line at 45° across the diagram represent those sites where Sitka spruce is growing faster than Western hemlock; conversely, for points below the line, Western hemlock is the faster species.

Regression data: metric yield classes
($y = a + b x$)

y	x	a	b	Correlation Coefficient
SS LYC	WH LYC	−2·29	1·03	0·78***
‡WH LYC	SS LYC	8·38	0·60	

‡ figure reversed.

Key to fig. 9 **Location and Details of Plots Comparing Growth Rates of Western Hemlock and Sitka Spruce**

Number on Diagram	Species	Forest Name	Compt. No.	Planting Year	Local Yield Class	Soil Type	Survey Stand No.
1	WH	Dartmoor,	10a	1927	16	Podzolised Brown Earth	40
	SS	Devon	8b	1926	14	,, ,, ,,	41
2	WH	Halwill,	62/4/1d	1938	20	Surface Water Gley	48
	SS	Devon & Cornwall	62/4/4e	1938	18	,, ,, ,,	49
3	WH	Quantock,	49/3/52	1928	26	Oligotrophic Surface Gley	64
	SS	Somerset	49/3/52	1928	28	,, ,, ,,	65
4	WH	Coed-y-Brenin,	S17–21	1941	16	Ochreous Brown Earth	123
	SS	Merioneth	S17–21	1941	16	,, ,, ,,	124
5	WH	Cynwyd,	54	1928	24	Ochreous Brown Earth	130
	SS	Merioneth	—	1928	16	Non-peaty Surface Water Gley	131
6	WH	Beddgelert (Deudraeth),	19d	1931	18	(Ochreous Brown Earth	133
	SS	Merioneth	20–21	1949	22	(with Superficial Gleying	136
7	WH	Radnor,	66	1927	14	Ochreous Brown Earth	167
	SS	Rads. & Hereford	67b	1927	16	,, ,, ,,	204
8	WH	Ratagan,	16f	1931	22	Brown Earth	523
	SS	Inverness & Ross	15d	1931	18	Seepage (flush) Gley	524
9	WH	Torrachilty (Lael),	61e	1938	18	Brown Earth	526
	SS	Ross	61d	1938	16	Mineral Gley	527
10	WH	Montreathmont,	19	1938	14	Groundwater Gley	635
	SS	Angus & Kincardine	18	1938	12	Imperfectly drained Brown Earth	636
11	WH	Monaughty,	80	1928	20	Brown Earth	638
	SS	Moray	81	1929	22	Lessivated Brown Earth	639
12	WH	Achaglachgach,	48	1941	20	Podzolised Brown Earth	701
	SS	Argyll	48	1941	16	Humus Iron Podzol	702
13	WH	Eredine,	86	1940	18	Ironpan Soil	711
	SS	Argyll	86	1940	12	Ironpan Soil with Humus 'B'	712
14	WH	Glenfinart,	21	1941	20	Ochreous Brown Earth	722
	SS	Argyll	16	1941	16	,, ,, ,,	721
15	WH	Knapdale,	13c	1943	22	Brown Earth	725
	SS	Argyll	13a	1943	24	,, ,,	726
16	WH	Kielder District,	1348	1939	12	Mineral Gley	914
	SS	Northumberland	1350	1939	10	Peaty Gley	915
17	WH	Grizedale,	119	1942	14	Ochreous Brown Earth	955
	SS	Lancs.	118	1942	8	,, ,, ,,	956
18	WH	Thornthwaite,	343a	1948	16	Mineral Gley	959
	SS	Cumberland	343b	1948	14	,, ,,	960

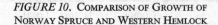

FIGURE 10. COMPARISON OF GROWTH OF
NORWAY SPRUCE AND WESTERN HEMLOCK

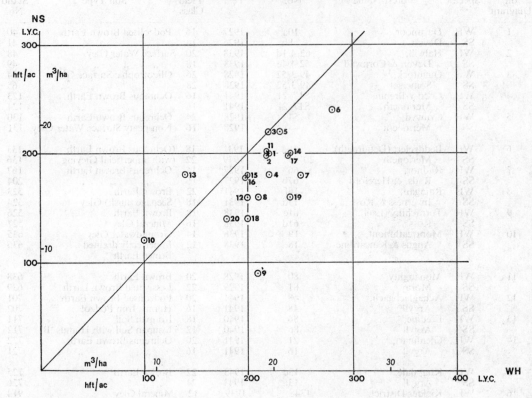

Each point represents a site where both species are growing. Points shown above the line at 45° across the diagram represent those sites where Norway spruce is growing faster than Western hemlock; conversely, for points below the line, Western hemlock is the faster species.

Regression data: metric yield classes
$(y = a + bx)$

y	x	a	b	Correlation Coefficient
NS LYC	WH LYC	5·84	0·53	0·57*
‡WH LYC	NS LYC	9·18	0·61	

‡ figure reversed.

Key to fig. 10 **Location and Details of Plots Comparing Growth Rates of Western Hemlock and Norway Spruce**

Number on Diagram	Species	Forest Name	Compt No.	Planting Year	Local Yield Class	Soil Type	Survey Stand No.
1	WH	Abinger (Chiddingfold),	67a	1943	18	Brown Gley	13
	NS	Surrey & Sussex	69c	1947	18	Non-peaty Surface Water Gley	17
2	WH	Abinger (Chiddingfold),	69a	1947	20	(Non Peaty	16
	NS	Surrey & Sussex	69e	1947	18	(Surface Water Gley	17
3	WH	Mynydd Du,	19d	1939	20	(Oligotrophic	95
	NS	Brecon & Mon.	19c	1939	20	(Brown Earth	96
4	WH	Talybont,	7a	1938	20	(Oligotrophic	106
	NS	Brecon	3d	1948	16	(Brown Earth	110
5	WH	Talybont,	39a	1943	20	(Oligotrophic	112
	NS	Brecon	38a	1943	20	(Brown Earth	113
6	WH	Tintern,	91d	1943	26	Podzolised Brown Earth	116
	NS	Monmouth	93a	1941	22	,, ,, ,,	117
7	WH	Coed-y-Brenin,	P13o	1942	22	Ochreous Brown Earth	121
	NS	Merioneth	P13i	1942	16	,, ,, ,,	122
8	WH	Beddgelert (Deudraeth),	19d	1931	18	Ochreous Brown Earth with	133
	NS	Merioneth	6a	1927	14	Superficial Gleying	137
9	WH	Bagley Wood,	15	1909	18	Podzolic Gley	172
	NS	Oxford	16	1907	8	Non-peaty Surface Water Gley	175
10	WH	Leanachan (Clunes),	30h	1937	8	Brown Earth	501
	NS	Inverness	30g	1937	10	Ochreous Brown Earth	502
11	WH	Leanachan (Clunes),	59g	1940	18	Brown Earth	506
	NS	Inverness	59a	1940	20	,, ,,	507
12	WH	Leanachan,	60	1939	18	Humus Iron Podzol	542
	NS	Inverness & Argyll	60	1939	14	Brown Earth with Surface Gleying	543
13	WH	Strathardle,	6	1931	12	Seepage Gley	606
	NS	Perth	6	1931	16	Mineral Gley	607
14	WH	Achaglachgach,	76	1939	22	Brown Earth	703
	NS	Argyll	76	1939	18	Brown Earth with Superficial Gleying	704
15	WH	Achaglachgach,	43	1941	18	Ochreous Brown Earth	705
	NS	Argyll	43	1941	16	,, ,, ,,	706
16	WH	Loch Ard,	83	1941	18	Ochreous Brown Earth	732
	NS	Perth & Stirling	83	1941	16	Slightly Ochreous Brown Earth	733
17	WH	Glenbranter,	52	1939	22	Ochreous Brown Earth	736
	NS	Argyll	54	1939	18	Brown Earth	737
18	WH	Inverliever (Inverinan),	—	1942	18	Ochreous Brown Earth	748
	NS	Argyll	—	1942	12	Ochreous Brown Earth with Superficial Gleying	749
19	WH	Queensberry Estate	—	1939	22	Podzolic Gley	822
	NS		—	1939	14	Brown Earth	823
20	WH	Allerston District,	SS25f	1943	16	Humus Iron Podzol	901
	NS	Yorks.	—	1946	12	,, ,, ,,	902

FIGURE 11. COMPARISON OF GROWTH OF DOUGLAS FIR AND WESTERN HEMLOCK

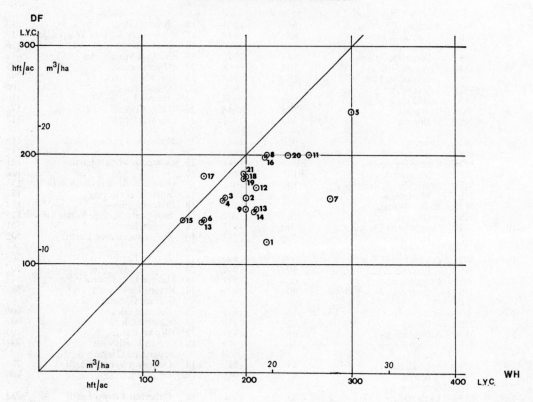

Each point represents a site where both species are growing. Points shown above the line at 45° across the diagram represent those sites where Douglas fir is growing faster than Western hemlock; conversely, for points below the line, Western hemlock is the faster species.

Regression data: metric yield classes
$(y = a + b x)$

y	x	a	b	Correlation Coefficient
DF LYC	WH LYC	7·26	0·43	0·60**
‡WH LYC	DF LYC	5·68	0·85	

‡ figure reversed.

Key to fig. 11 **Location and Details of Plots Comparing Growth Rates of Western Hemlock and Douglas Fir**

Number on Diagram	Species	Forest Name	Compt. No.	Planting Year	Local Yield Class	Soil Type	Survey Stand No.
1	WH	Alice Holt,	38a	1947	20	(Non-peaty Surface	3
	DF	Hants & Surrey	25b	1944	10	(Water Gley	6
2	WH	Andover,	8	1950	18	Lessivated Brown Earth	9
	DF	Hants	8	1950	14	,, ,, ,,	10
3	WH	Cranborne Chase,	12	1943	16	Calcareous Brown Earth	18
	DF	Dorset & Wilts	17	1942	14	Lessivated Brown Earth	19
4	WH	New Forest,	1/39a	1941	16	Brown Earth	22
	DF	Hants	20c	1920	14	Humus Iron Podzol	25
5	WH	Dartington Estate,	I	1933	28	Oligotrophic Brown Earth	35
	DF	Devon	I	1934	22	Eutrophic Brown Earth	36
6	WH	Quantoxhead Estate,	—	1900	14	Iron Podzol	46
	DF	Somerset	—	1900	12	,, ,,	47
7	WH	Dartmoor (Okehampton),	64/4/14d	1948	24	Brown Gley	54
	DF	Devon	64/4/14b	1944	14	Ochreous Brown Gley	55
8	WH	Dartmoor (Plym),	15	1951	20	Ochreous Brown Earth	61
	DF	Devon	7	1954	18	,, ,, ,,	62
9	WH	Tair Onen,	11c	1925	18	Podzolised Brown Earth	89
	DF	Glamorgan	11c	1923	14	,, ,, ,,	90
10	WH	Beddgelert (Deudraeth),	19d	1931	18	(Ochreous Brown Earth	133
	DF	Merioneth.	19c	1931	14	(with Superficial Gleying	134
11	WH	Dovey (Corris),	41a	1928	24	Ochreous Brown Earth	139
	DF	Merioneth & Mont.	41b	1928	18	Ochreous Brown Earth with Superficial Gleying	140
12	WH	Gwydyr,	320	1927	18	Ochreous Brown Earth	157
	DF	Caerns. & Denbs.	321h	1921	16	,, ,, ,,	158
13	WH	Radnor,	66	1927	14	Ochreous Brown Earth	167
	DF	Radnor & Hereford	66d	1927	12	,, ,, ,,	168
14	WH	Bagley Wood,	15	1909	18	Podzolic Gley	172
	DF	Oxford	16	1907	14	Non-peaty Surface Water Gley	174
15	WH	Hereford Forest (Dymock),	21	1915	12	Brown Gley	198
	DF	Glos. & Hereford	9b	1921	12	Non-peaty Surface Water Gley	199
16	WH	Glen Urquhart,	17a	1933	20	Brown Earth	508
	DF	Inverness	17b	1933	18	Mineral Gley	509
17	WH	Inchnacardoch	14a	1934	14	Ochreous Brown Earth	528
	DF	(Portclair), Inverness	14b	1934	16	,, ,, ,,	529
18	WH	Black Isle,	18	1928	18	Podzolic Brown Earth	608
	DF	Ross	18	1928	16	Iron Podzol	609
19	WH	Culloden,	25c	1927	18	Podzolised Brown Earth	616
	DF	Inverness & Nairn	25b	1927	16	,, ,, ,,	617
20	WH	Solway (Dalbeattie),	415	1926	22	Brown Earth	801
	DF	Kirkcudbright	415	1926	18	,, ,,	802
21	WH	Dunnerdale,	6	1942	18	Brown Earth	953
	DF	Cumbs & Lancs	6	1942	16	,, ,,	954

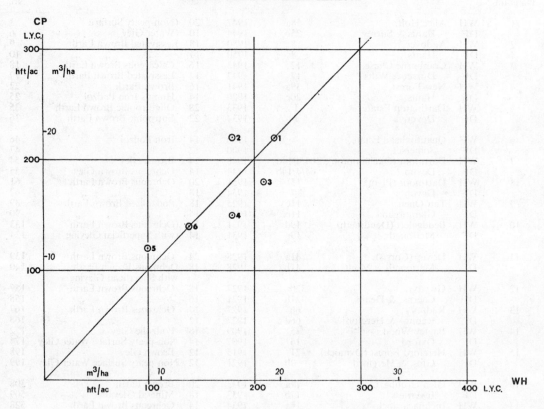

FIGURE 12. COMPARISON OF GROWTH OF CORSICAN PINE AND WESTERN HEMLOCK

Each point represents a site where both species are growing. Points shown above the line at 45° across the diagram represent those sites where Corsican pine is growing faster than Western hemlock; conversely, for points below the line, Western hemlock is the faster species.

Regression data: metric yield classes
$(y = a + bx)$

y	x	a	b	Correlation Coefficient
CP LYC	WH LYC	3·79	0·75	0·80***
‡WH LYC	CP LYC	2·24	0·85	

‡ figure reversed.

Key to fig. 12 **Location and Details of Plots Comparing Growth Rates of Corsican Pine and Western Hemlock**

Number on Diagram	Species	Forest Name	Compt. No.	Planting Year	Local Yield Class	Soil Type	Survey Stand No.
1	WH	Alice Holt,	38a	1947	20	Surface Water Gley	3
	CP	Hants	27a	1945	20	,, ,, ,,	4
2	WH	New Forest (Broomy)	1/39a	1941	16	Brown Earth	22
	CP	New Forest (Rhinefield), (Hants)	40c	1946	18	Iron Podzol	23
3	WH	Gwydyr,	320	1927	18	Ochreous Brown Earth	157
	CP	Caerns.	321	1931	16	,, ,, ,,	159
4	WH	Thetford,	157	1951	16	Podzolised Brown Earth	228
	CP	Norfolk & Suffolk	158	1927	14	,, ,, ,,	229
5	WH	North Lindsey	17c	1941	8	Humus Iron Podzol	235
	CP	(Willingham), Lincs.	21	1950	10	,, ,, ,,	237
6	WH	Kesteven,	28	1931	12	Humic Gley	245
	CP	Lincs & Rutland	28	1931	12	,, ,,	246

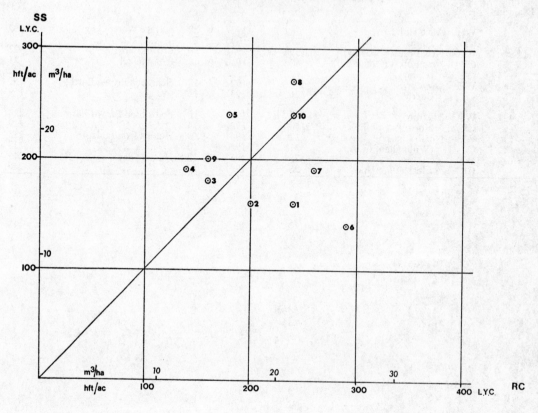

FIGURE 13. COMPARISON OF GROWTH OF SITKA SPRUCE AND WESTERN RED CEDAR

Each point represents a site where both species are growing. Points shown above the line at 45° across the diagram represent those sites where Sitka spruce is growing faster than Western red cedar; conversely, for points below the line, Western red cedar is the faster species.

Regression data: metric yield classes
$(y = a + bx)$

y		x		a	b	Correlation Coefficient
SS	LYC	RC	LYC	19·4	−0·10	−0·12
‡RC	LYC	SS	LYC	21·4	−0·14	

‡ figure reversed.

Key to fig. 13 **Location and Details of Plots Comparing Growth Rates of Sitka Spruce and Red Cedar**

Number on Diagram	Species	Forest Name	Compt. No.	Planting Year	Local Yield Class	Soil Type	Survey Stand No.
1	RC	Eggesford,	25/27	1920	22	Brown Gley	42
	SS	Devon	25/27	1920	14	Non-peaty Surface Water Gley	43
2	RC	Eggesford,	2a	1921	18	Non-peaty Surface Water Gley	44
	SS	Devon	25/27	1920	14	Ochreous Brown Earth	43
3	RC	Crychan,	24f	1936	14	Ochreous Brown Earth	86
	SS	Brecon & Carms.	25b	1935	16	,, ,, ,,	84
4	RC	Mynydd Du,	2c	1935	12	Oligotrophic Brown Earth	97
	SS	Brecon & Mon.	2a	1935	18	,, ,, ,,	98
5	RC	Beddgelert (Deudraeth),	19j	1939	16	Non-peaty Surface Water Gley	135
	SS	Merioneth.	20–21	1949	22	Ochreous Brown Earth with Superficial Gleying	136
6	RC	Dyfi (Corris),	30b	1932	26	(Ochreous Brown Earth with	143
	SS	Merioneth & Mont.	95b	1935	12	(Superficial Gleying	142
7	RC	Gwydyr,	164k	1930	24	Ochreous Brown Earth	149
	SS	Caerns. & Denbs.	322a	1924	18	,, ,, ,,	150
8	RC	Vyrnwy Estate,	5	1951	22	(Ochreous Brown Earth with	169
	SS	Montgomery	5	1951	24	(Superficial Gleying	170
9	RC	Ratagan, Ross	16e	1931	14	Brown Earth	525
	SS		15d	1931	18	Seepage (flush) Gley	524
10	RC	Laigh of Moray	80	1928	22	Ochreous Brown Earth	640
	SS	(Monaughty), Moray	81	1929	22	Lessivated Brown Earth	639

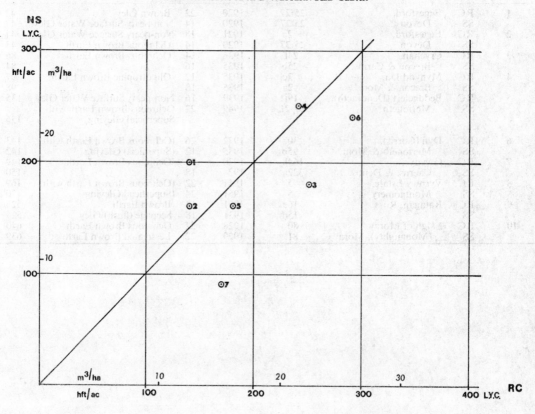

FIGURE 14. COMPARISON OF GROWTH OF NORWAY SPRUCE AND WESTERN RED CEDAR

Each point represents a site where both species are growing. Points shown above the line at 45° across the diagram represent those sites where Norway spruce is growing faster than Western red cedar; conversely, for points below the line, Western red cedar is the faster species.

Regression data: metric yield classes
$(y = a + bx)$

y		x		a	b	Correlation Coefficient
NS	LYC	RC	LYC	6·51	0·55	0·59
‡RC	LYC	NS	LYC	7·60	0·64	

‡ figure reversed.

Key to fig. 14 **Location and Details of Plots Comparing Growth Rates of Norway Spruce and Red Cedar**

Number on Diagram	Species	Forest Name	Compt. No.	Planting Year	Local Yield Class	Soil Type	Survey Stand No.
1	RC	Abinger (Chiddingfold),	68	1943	12	(Non-peaty Surface Water	14
	NS	Surrey & Sussex	69c	1947	18	(Gley	17
2	RC	Caeo,	33e	1939	16	(Non-peaty Surface Water	82
	NS	Carmarthen	33a	1939	14	(Gley	80
3	RC	Talybont	7f	1939	20	(Oligotrophic Brown	107
	NS	Brecon	3d	1948	16	(Earth	110
4	RC	Wentwood,	1a	1946	22	(Oligotrophic Brown	118
	NS	Monmouth	1c	1946	22	(Earth	120
5	RC	Beddgelert (Deudraeth),	19j	1939	16	Non-peaty Surface Water Gley	135
	NS	Merioneth	6a	1937	14	Ochreous Brown Earth with Superficial Gleying	137
6	RC	Dyfi (Corris),	30b	1932	26	(Ochreous Brown Earth with	143
	NS	Merioneth & Mont.	30a	1933	22	(Superficial Gleying	144
7	RC	Bagley Wood,	15	1909	16	Podzolic Gley	171
	NS	Oxford	16	1907	8	Non-peaty Surface Water Gley	175

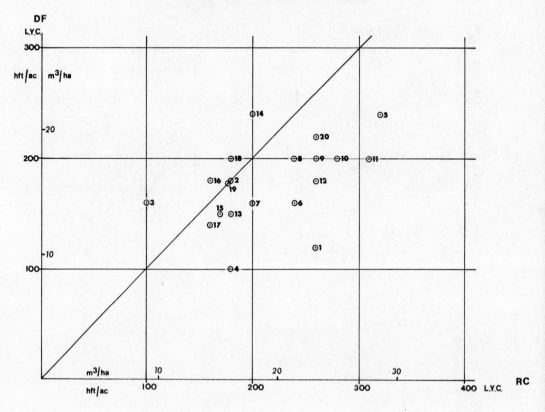

FIGURE 15. COMPARISON OF GROWTH OF DOUGLAS FIR AND WESTERN RED CEDAR

Each point represents a site where both species are growing. Points shown above the line at 45° across the diagram represent those sites where Douglas fir is growing faster than Western red cedar; conversely, for points below the line, Western red cedar is the faster species.

Regression data: metric yield classes
$(y = a + b\,x)$

y	x	a	b	Correlation Coefficient
DF LYC	RC LYC	10·2	0·29	0·46*
‡RC LYC	DF LYC	7·92	0·71	

‡ figure reversed.

Key to fig. 15 **Location and Details of Plots Comparing Growth Rates of Douglas Fir and Red Cedar**

Number on Diagram	Species	Forest Name	Compt. No.	Planting Year	Local Yield Class	Soil Type	Survey Stand No.
1	RC	Alice Holt,	29i	1945	24	(Non-peaty Surface	5
	DF	Hants & Surrey	25b	1944	10	(Water Gley	6
2	RC	Alice Holt,	53c	1906	16	(Non-peaty Surface Water	7
	DF	Hants & Surrey	53e	1906	16	(Gley	8
3	RC	New Forest,	21g	1919	8	Humus Iron Podzol	24
	DF	Hants	20c	1920	14	,, ,, ,,	25
4	RC	Challock (Orlestone),	7e	1944	16	(Non-peaty Surface Water	26
	DF	Kent	7c	1944	8	(Gley	27
5	RC	Dartington Estate,	VI	1937	28	Brown Gley	38
	DF	Devon	IV	1942	22	Eutrophic Brown Earth	39
6	RC	Eggesford,	25/27	1920	22	Brown Gley	42
	DF	Devon	1	1921	14	Ochreous Brown Earth	45
7	RC	Eggesford,	2a	1921	18	Ochreous Brown Earth	44
	DF	Devon	1	1921	14	,, ,, ,,	45
8	RC	Dartmoor (Plym),	15	1951	22	,, ,, ,,	59
	DF	Devon	7	1954	18	,, ,, ,,	62
9	RC	Quantock,	49/3/3d	1923	24	Oligotrophic Brown Earth	68
	DF	Somerset	49/3/4	1924	18	,, ,, ,,	69
10	RC	Tavistock Woodlands Ltd.,	D2J	1910	24	Ochreous Brown Earth	70
	DF	Devon	DsF	1907	18	,, ,, ,,	71
11	RC	Tavistock Woodlands Ltd.,	D2M	1950	72	Ochreous Brown Earth	72
	DF	Devon	DsF	1907	18	,, ,, ,,	71
12	RC	Coed-y-Brenin,	R15c	1931	24	,, ,, ,,	126
	DF	Merioneth	R15a	1927	16	,, ,, ,,	127
13	RC	Beddgelert (Deudraeth),	19j	1939	16	(Ochreous Brown Earth with	135
	DF	Merioneth	19c	1931	14	(Superficial Gleying	134
14	RC	Gwydyr,	331i	1930	18	Eutrophic Brown Earth	160
	DF	Caerns. & Debs.	321	1927	22	Ochreous Brown Earth	162
15	RC	Bagley Wood,	15	1909	16	Podzolic Gley	171
	DF	Oxford	16	1907	14	Non-peaty Surface Water Gley	174
16	RC	Dean,	79b	1916	14	Podzolised Brown Earth	193
	DF	Glos., Hereford, Mon.	79e	1929	18	Oligotrophic Brown Earth	194
17	RC	Hereford Forest (Dymock),	28	1914	14	(Non-peaty Surface Water	197
	DF	Glos. & Hereford)	9b	1921	12	(Gley	199
18	RC	Glen Urquhart,	36d	1930	16	Ochreous Brown Earth	510
	DF	Inverness	33a/b	1930	18	,, ,, ,,	511
19	RC	Inchnacardoch	14e	1934	16	Brown Earth	530
	DF	(Portclair), Inverness	14b	1934	16	Ochreous Brown Earth	529
20	RC	Culloden,	201	1927	24	Brown Earth	618
	DF	Inverness & Nairn	201	1927	20	,, ,,	619

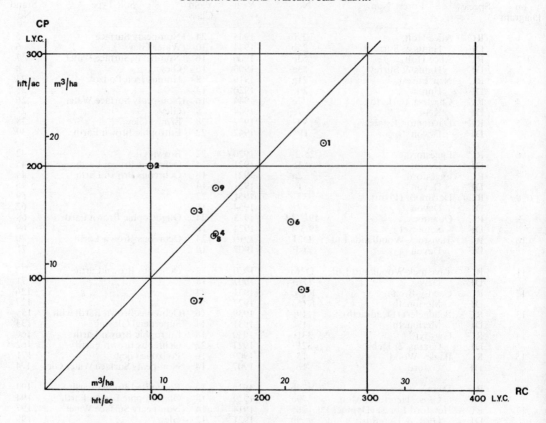

FIGURE 16. COMPARISON OF GROWTH OF
CORSICAN PINE AND WESTERN RED CEDAR

Each point represents a site where both species are growing. Points shown above the line at 45° across the diagram represent those sites where Corsican pine is growing faster than Western red cedar; conversely, for points below the line, Western red cedar is the faster species.

Regression data: metric yield classes
($y = a + b x$)

y	x	a	b	Correlation Coefficient
CP LYC	RC LYC	13·3	0·01	0·01
‡RC LYC	CP LYC	15·6	0·01	

‡ figure reversed.

Key to fig. 16 **Location and Details of Plots Comparing Growth Rates of Corsican Pine and Red Cedar**

Number on Diagram	Species	Forest Name	Compt. No.	Planting Year	Local Yield Class	Soil Type	Survey Stand No.
1	RC	Alice Holt,	29i	1945	24	(Non-peaty Surface Water	5
	CP	Hants & Surrey	27a	1945	20	(Gley	4
2	RC	New Forest,	21g	1919	8	Humus Iron Podzol	24
	CP	Hants	40c	1946	18	Iron Podzol	23
3	RC	Ebbw (Draethen),	16b	1920	12	(Oligotrophic Brown Earth	87
	CP	Glam. & Mon.	16a	1920	14	,, ,, ,,	88
4	RC	Rockingham,	23b	1921	14	Calcareous Brown Earth	200
	CP	Northants & Hunts	22d	1916	12	Brown Gley	201
5	RC	Ampthill,	2	1950	22	(Non-peaty Surface Water	221
	CP	Beds. & Hants.	5	1951	8	(Gley	223
6	RC	Thetford,	157	1951	16	Podzolised Brown Earth	228
	CP	Norfolk & Suffolk	158	1927	14	,, ,, ,,	229
7	RC	Kesteven,	78c	1950	12	(Non-peaty Surface Water	238
	CP	Lincs. & Rutland	78b	1950	8	(Gley	239
8	RC	North Lindsey	28	1931	14	Podzolised Brown Earth	243
	CP	(Laughton), Lincs.	28	1931	12	,, ,, ,,	244
9	RC	Ford Estate,	(Fenwick	1910	24	Brown Earth	906
	CP	Berwick	Wood)	1910	16	,, ,,	907

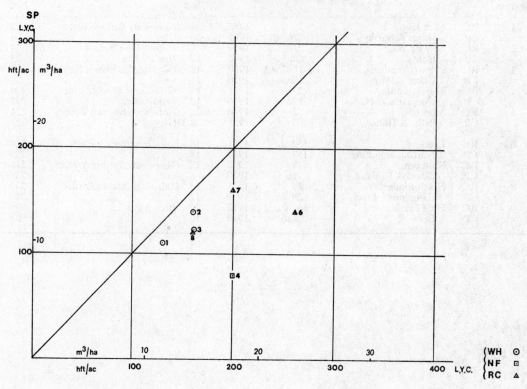

FIGURE 17. COMPARISON OF GROWTH OF SCOTS PINE AND THREE MINOR SPECIES

Each point represents a site where both species are growing. Points shown above the line at 45° across the diagram represent those sites where Scots pine is growing faster than three minor species; conversely, for points below the line, the three minor species grow faster.

Regression data:

y	x	a	b	Correlation Coefficient
LYC	LYC	insufficient		data
LYC	LYC			

Plate 1. Foliage on lower side branch of Grand fir.

Plate 3. 28-year-old Grand fir stems at Gwydyr Forest, Caernarvonshire. Local Yield Class 25. C 3262.

Plate 4. 24-year-old Western hemlock stand in the Corris block of Dyfi Forest, Merioneth. Local Yield Class 25. This stand is one of the best hemlock stands in Britain.   D 2446.

Plate 5. 50-year-old Western hemlock stems with natural regeneration in the background. Langridge Wood, Dunster, Somerset. B 1266.

Plate 7. Bark, vertical and transverse sections of Noble fir stem, showing drought crack and associated staining. CS 18095.

Plate 8 *above:* Grand fir.

Plates 8 to 11. PLANKS OF THE FOUR SPECIES. EACH GROUP OF PLANKS HAS BEEN SELECTED FROM MATERIAL TESTED AT PRINCES RISBOROUGH LABORATORY IN 1971/72, TO SHOW THE RANGE OF WOOD QUALITY FOUND. NOTE ESPECIALLY KNOTS, AND RATES OF GROWTH AS SHOWN BY END GRAIN.

Plate 9 *below:* Noble fir.

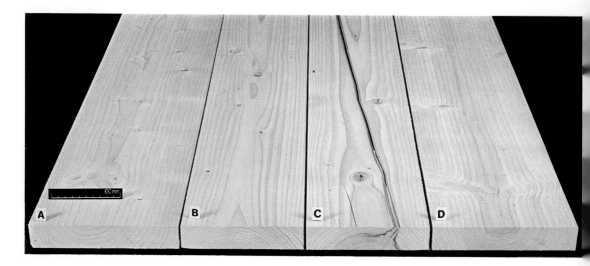

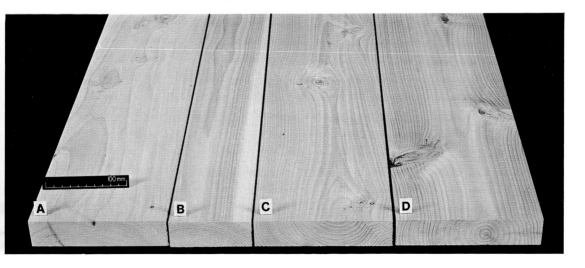

Plate 10 *above:* Western red cedar.

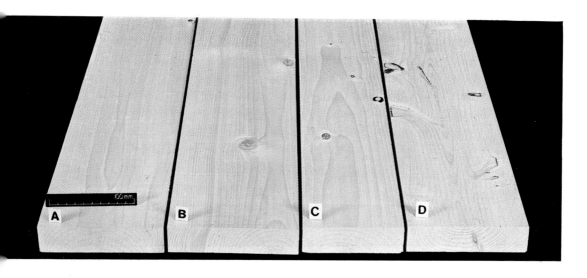

Plate 11 *below:* Western hemlock.

Plate 12. 13-year-old Western hemlock and Sitka spruce, Margam block of Coed Morgannwg, Glamorgan. 350 m elevation. Many of the hemlock on this moderately exposed site have multiple leaders.   CS 24349.

Plate 13. Stem of specimen Western red cedar showing fluting. (This is more extremely developed than normal). Eggesford Forest, Devon.   D 4292.

Plate 14. 28-year-old Western red cedar stand, Queen Elizabeth Forest, Hants and Sussex. This stand is growing on a shallow brown earth over chalk on the top of Wardown. Height at 25 years, 10.7 m.

Plate 15. A 26-year-old plantation of Western red cedar, Alice Holt Forest, Hampshire. Local Yield Class 13. C 2154.

Plate 16. 17-year-old Noble fir plot, Kilmun Forest Garden, near Dunoon, Argyll. Mean height 5.9 m at 15 years.

Plate 17. 28-year-old Noble fir stems, showing severe drought crack, Gwydyr Forest, Caernarvonshire. Local Yield Class 19. C 3051.

Plate 18. Noble fir specimen tree, Bedgebury Pinetum, Kent.   B 611.

Key to fig. 17 **Location and Details of Plots Comparing Growth Rates of Scots Pine and Three Minor Species**

Number on Diagram	Species	Forest Name	Compt. No.	Planting Year	Local Yield Class	Soil Type	Survey Stand No.
1	WH	North Lindsey	17c	1941	8	Humus Iron Podzol	235
	SP	(Willingham), Lincs.	17a	1941	10	,, ,, ,,	236
2	WH	Ampleforth	196	1936	16	Humus Iron Podzol	908
	SP	(Hambleton), York	—	1936	12	,, ,, ,,	909
3	WH	Hamsterley,	1a	1941	12	Brown Earth	912
	SP	Durham	1a	1941	12	Slightly Podzolic Brown Earth	913
4	NF	Alltcailleach,	18	1929	18	Brown Earth	603
	SP	Aberdeen	21	1923	8	Imperfectly Drained Brown Earth	604
5	RC	Rockingham,	22	1952	20	(Non-peaty Surface Water	203
	SP	Northants. & Hunts.	11	1948	12	(Gley	202
6	RC	Darnaway Estate,	6	1914	20	Ochreous Brown Earth	620
	SP	Morayshire	6	1914	12	Brown Earth	621
7	RC	Dechant Estate,	4	1920	16	Podzolic Brown Earth	904
	SP	Naboth	4	1920	10	,, ,, ,,	905

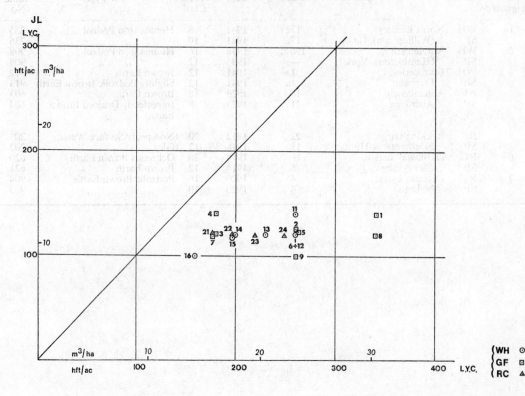

FIGURE 18. COMPARISON OF GROWTH OF
JAPANESE LARCH AND THREE MINOR SPECIES

Each point represents a site where both species are growing. Points shown above the line at 45° across the diagram represent those sites where Japanese larch is growing faster than three minor species; conversely, for points below the line, the three minor species grow faster.

Regression data:

y	x	a	b	Correlation Coefficient
LYC	LYC	insufficient		data
LYC	LYC			

GROWTH IN OLDER CROPS

Key to fig. 18 **Location and Details of Plots Comparing Growth Rates of Japanese Larch and Three Minor Species**

Number on Diagram	Species	Forest Name	Compt No.	Planting Year	Local Yield Class	Soil Type	Survey Stand No.
1	GF	Bodmin,	43/2/5c	1951	30	Ochreous Brown Earth	33
	JL	Cornwall	43/2/5a	1951	10	,, ,, ,,	34
2	GF	Caeo,	46d	1943	24	(Ochreous Brown Earth with	78
	JL	Carmarthen	46a	1940	8	(Superficial Gleying	79
3	GF	Crychan,	25a	1937	16	(Ochreous Brown Earth with	83
		Brecon. & Carms.				(Superficial Gleying	
	JL		24a	1953	8	Ochreous Brown Earth	85
4	GF	Coed Abertawe,	3b	1933	18	(Non-peaty Surface Water	105
	JL	Glamorgan	3a	1934	10	(Gley	104
5	GF	Talybont,	3c	1942	24	Oligotrophic Brown Earth	108
	JL	Brecon	8b	1938	10	,, ,, ,,	111
6	GF	Talybont,	4c	1942	22	,, ,, ,,	109
	JL	Brecon	8b	1938	10	,, ,, ,,	111
7	GF	Bagley Wood,	25e	1904	16	Non-peaty Surface Water Gley	176
	JL	Oxford	15	1909	12	Podzolic Gley	173
8	GF	Fleet,	22	1946	34	Brown Earth	805
	JL	Kirkcudbright	22	1946	12	,, ,,	806
9	GF	Kirroughtree,	39	1940	24	Brown Earth	815
	JL	Kirkcudbright	39	1940	6	,, ,,	816
11	WH	Bodmin,	43/2/5d	1951	22	Ochreous Brown Earth	32
	JL	Cornwall	43/2/5a	1951	10	,, ,, ,,	34
12	WH	Brechfa (III),	472b	1938	24	,, ,, ,,	73
	JL	Carmarthen	472a	1938	10	,, ,, ,,	74
13	WH	Caeo,	46c	1943	20	(Ochreous Brown Earth with	77
	JL	Carmarthen	46a	1940	8	(Superficial Gleying	79
14	WH	Talybont,	7a	1938	20	Oligotrophic Brown Earth	106
	JL	Brecon	8b	1938	10	with Superficial Gleying	111
15	WH	Bagley Wood,	15	1909	18	Podzolic Gley	172
	JL	Oxford	15	1909	12	,, ,,	173
16	WH	Solway (Dalbeattie),	77	1941	14	Brown Earth	803
	JL	Kirkcudbright	77	1941	8	,, ,,	804
21	RC	Crychan,	24f	1936	14	Ochreous Brown Earth	86
	JL	Brecon & Carms.	24a	1935	8	,, ,, ,,	85
22	RC	Talybont,	7f	1939	20	Oligotrophic Brown Earth	107
	JL	Brecon	8b	1938	10	,, ,, ,,	111
23	RC	Bagley Wood,	15	1909	16	Podzolic Gley	171
	JL	Oxford	15	1909	12	,, ,,	173
24	RC	Inverliever,	75	1914	22	(Brown Earth with	723
	JL	Argyll	75	1914	10	(Superficial Gleying	724

3.1 continued. Rates of Growth in Pole-stage and Older Crops

The relationships between pairs of species, as shown in Table 9 and Figs. 1–18 can be grouped as follows:

(a) Those where the minor species becomes relatively more productive on higher yielding sites.

(b) Those where the minor species has a consistent advantage over the major species for all sites observed, even if this advantage is small.

(c) Those where no consistent relationship between species is apparent in the data as presented, either because points are widely scattered, or because the data are insufficient to support any conclusions.

The first category, where the minor species are at a greater advantage the better the site, includes the most important Grand fir comparisons, namely Sitka spruce/Grand fir; Douglas fir/Grand fir; Corsican pine/Grand fir; and also Douglas fir/Red cedar and Corsican pine/Western hemlock. For all pairs of species in this group, it is implicit that on intermediate sites the rate of growth of both species may be equal and on poor sites the major species may out-yield the minor species.

The second category, where the minor species has a consistent, though sometimes small advantage in volume production over the major species, includes most comparisons of hemlock, together with the Norway spruce/Grand fir and Sitka spruce/Noble fir comparisons. In this group, the greatest difference in growth rates is between Grand fir and Norway spruce, Grand fir producing between two and four yield classes more than Norway spruce on most sites. Hemlock generally yields between one and three yield classes more than either of the spruces or Douglas fir.

The last category, i.e. where the relationship between species is inconsistent or indeterminate includes three of the four comparisons involving Western red cedar, and three of the four involving Noble fir. The available data shows Red cedar to be at its best on lowland sites and to have a greater advantage over other species on imperfectly drained soils than on those with better drainage. In Scotland, Red cedar has less of an advantage over major species than in England and Wales.

There is no consistent relationship between Noble fir and Norway spruce; the data comparing Noble fir both with Douglas fir and Corsican pine are inadequate.

On chalk downland in the south of England, most conifers have been notably unsuccessful, many having been afflicted with lime-induced chlorosis after making promising early growth (Wood and Nimmo, 1962). Western red cedar remains one of the few coniferous species usable on such sites.

3.11 Limitations of the Data on Growth Rates

While the results presented in the preceding paragraph and accompanying diagrams appear to give some consistent relationships, the stands from which the data were obtained are limited in age range and site and may, therefore, give rise to biased results.

The absence of older stands has resulted in selection of the bulk of surveyed plots in stands 20 to 40 years old. While it is to be hoped that in the second half of the rotation all species will continue to grow satisfactorily until felling is due, this has not been demonstrated in any more than a handful of stands. In arboreta, growth of minor species has mostly continued to an age well past the present rotation age of between 50 and 60 years. However, for Grand fir, there are indications of premature death or die-back on some exposed or dry sites giving rise to a doubt about the future of younger stands in similar situations. Die-back of Grand fir and its relationship to *Adelges piceae* and exposure are discussed in Section 5.41.

Lastly, the limitations of the data must not be overlooked. The plots included in the survey of minor species were assessed by skilled and experienced staff, but the sample of each of the characteristics assessed was reduced to the bare minimum. This was most marked in the assessment of yield class, which was based on the height of the six trees of largest girth in a 0.15 acre (0.06 hectare) plot. The Production Class (or Local Yield Class) was obtained by modifying the General Yield Class according to the basal area of the plot or the mean girth of the dominant trees.

3.2 Stand Structure and Thinning

The four minor species contrast markedly in stand structure. Western hemlock and Red cedar stands tend to be uniform and to have relatively small range in height and girth (similar in this respect to larches or Corsican pine). In contrast Noble fir and, to a lesser degree, Grand fir stands can include a very wide range of stem sizes mainly because of the shade tolerance of sub-dominant and suppressed trees. These can survive under conditions of heavy shade where similar trees of other species would die. The range of height and girth of Noble fir stems is greatest on the more difficult sites where groups or individuals may get away several years before the remainder of the crop. The same pattern of growth can be observed in other species on similar sites but it is more pronounced with Noble fir than any other species.

Red cedar stands are exceptional in that there is little interpenetration of branches from the crown of one tree to the next. Even so, crowns are quite readily restricted by shading and by competition from adjacent trees.

TABLE 10
COMPARISON OF PLANTATIONS WHEN AT 20 M TOP HEIGHT (YC 14)

Species	Stems per Hectare	Mean Diam. cm	Basal Area sq m per ha	Average Vol. per Tree Thinned cu m	Age at which Height reached	M A I at 20 m (Yield Class 14) cu. m. per hectare
Western hemlock	894	21	31·8	0·23	44	12·4
Red cedar	1073	24	49·3	0·28	50	13·1
Grand fir	795	22	31·0	0·24	39	12·1
Noble fir	997	23	42·4	0·30	50	13·1
Sitka spruce	694	24	25·7	0·30	43	13·0
Norway spruce	722	24	33·5	0·28	46	12·3
Douglas fir	626	23	25·7	0·25	34	11·7
Corsican pine	545	27	30·7	0·45	43	13·5
Japanese larch	374	26	20·7	0·46	31	13·5

Source: Forestry Commission Booklet 34: *Forest Management Tables* (*Metric*).

3.21 First Thinning

For any given yield class, the first thinning of stands of minor species occurs later, according to Forestry Commission Management Tables (Hamilton and Christie, 1971), than thinnings in stands of almost all the major species of similar yield class. Hemlock and Grand fir may become ready for first thinning three or four years later than Sitka spruce or Douglas fir stands of the same age and yield class, while Red cedar and Noble fir become due for first thinning two to four years later still.

Thinning of Grand fir and Noble fir is delayed because of their initially slow growth so that canopy closes and thicket is formed several years later than species which get away quickly. In Western red cedar stands, thinnings are probably deferred because the crowns appear to be free and are slow to respond to thinning. In contrast, thinning in hemlock is delayed because of its form. The species is slender for its height; to obtain produce of a given minimum volume or diameter, first thinning has to be deferred beyond the time when it would be economic to thin stands of other species of the same height.

Subsequently, when given standard 'Management Table' thinnings, all four minor species for a given height carry a larger number of stems per hectare. Table 10 illustrates that for Western hemlock and Grand fir, both of which reach 25 m at approximately the same age as Sitka spruce, the volume and girth of trees removed in thinning is appreciably less than for Sitka spruce.

Noble fir and Red cedar thinnings are similar in size to Sitka spruce thinnings at the same height but the Noble fir and cedar stands reach this height 5 or more years later than almost all other species. Thus, for a given yield class, whether because thinnings are smaller in volume or are harvested later, the Discounted Revenue for the minor species is bound to be less than for the major species.

(See also Section 3.4, page 51, on crop stability.)

3.3 Defects Due to Adverse Climatic and Site Factors

Table 11 summarises the observations made on the incidence of drought crack, forking, poor stem form, fluting, buttressing, crown blast and windthrow observed during the main survey of minor species.

TABLE 11

STEM DEFECTS AND WINDTHROW INCIDENCE OBSERVED DURING SURVEY OF STANDS

Defect	Western hemlock	Red cedar	Grand fir	Noble fir	Sitka spruce	Norway spruce	Douglas fir	Corsican pine	Larches*
No. of stands observed	72	48	67	26	44	39	46	20	17
Drought Crack Incidence %	11	0	36	62	2	5	2	0	0
Severity Index	1·1	—	1·8	3·0	1·0	1·5	1·0	—	—
Forked stems/Double leaders Incidence %	29	30	1	0	2	5	0	0	0
Severity Index	1·6	1·9	1·0	—	1·0	1·0	—	—	—
Poor stem form (Wavy stems, spiral grain, butt sweep) Incidence %	17	6	4	19	0	5	9	5	65
Severity Index	2·4	1·3	1·0	1·0		1·0	1·5	1·5	1·8
Fluting Incidence %	46	31	10	19	0	15	0	0	0
Severity Index	1·7	1·6	1·0	1·4	—	1·2	—	—	—
Buttressing/butt swelling Incidence %	1	18	0	0	0	0	0	0	0
Windthrow Incidence %	31	26	43	58	48	28	46	15	12
Score	1·0	1·0	1·0	1·0	1·2	1·0	1·2	1·0	1·0
Crown blast/ill health Incidence %	8	4	12	0	0	2	11	5	0
Severity Index	1·2	1·0	1·4	—	—	0	1·2	1·0	—

*The 17 larch plots consist of 11 Japanese, 4 European and 2 Hybrid larch plots. The incidence of defects was similar for both species and the hybrid which are therefore considered together.

See page 49, facing, for *Footnote*.

Footnote to Table 11

The "incidence %" and "severity index" values in the table summarise the observations made. The "incidence %" is the proportion of the visited stands in which a defect was seen; it takes no account of the extent of a defect in any stand. The extent of a defect is usually given by a "severity index" and in one instance by an "incidence score". In all stands where a defect was observed, frequency and severity were each independently scored on a 1 to 3 scale. The "severity index" was obtained by multiplying the mean frequency score in affected plots of a species by the mean severity score. The "severity index" figures can in theory range from 1.0 (light damage on less than a third of the trees in a stand) to 9.0 (severe damage on more than two-thirds of the crop). In the table, it will be seen that there are few index values above 2.0 and none above 3.0.

3.31 Drought Crack

Stem cracking, associated with late season droughts and high summer temperatures, has been reported in Britain for a number of coniferous species including Sitka spruce, Norway spruce, Western hemlock, Grand fir and Noble fir (Day, 1954). Drought cracks vary in length from a few centimetres up to 9 metres or more; they may be 5 to 8 centimetres deep and up to 2 centimetres wide. After one to two years, new wood usually grows over the crack and normal wood formation is resumed, but the scar on the bark persists and is the usual diagnostic feature of drought crack (Plates 12, 15).

Western hemlock: drought crack

Out of 72 stands examined, 8 (i.e. 11 per cent) had stem crack scars present and in each case fewer than one-third of the trees were affected. The severity of scars was rated as slight in 7 of the affected stands and as moderate in the eighth. Scarring was associated with increasing crop age and tree size, rather than any site factor.

No information is available about the relationship between external scarring and internal timber cracking in Western hemlock. However, in view of the infrequency and low severity rating of external scarring, drought crack is unlikely to be of importance in relation to the utilisation of Western hemlock timber.

Western red cedar: drought crack

None of the 48 stands examined included trees with noticeable bark scarring.

Grand fir: drought crack

Noticeable external bark scarring was reported for 24 (36 per cent) of the 67 stands examined. Of the affected stands, 20 had fewer than one-third of their trees with scarring, while in the remaining 4 between one-third and two-thirds of the trees were scarred. The severity of scarring was slight in 15 of the stands and moderate in the other 7. The occurrence of scarring was associated with increasing crop age and size of trees, poor soil drainage and shallow rooting depth. In general, these data are in agreement with earlier reports that external scarring of Grand fir was of relatively wide-spread occurrence.

Fears that the incidence of drought crack associated with the scarring would seriously reduce the value of Grand fir timber led to a detailed investigation in 1964–65 (Greig, 1969). In this, the incidence of external bark scarring and the relationship between scarring and internal crack were studied throughout Britain, together with the effect of stem crack on sawn timber out-turn. It was found that in the 102 stands examined, the proportion of trees with scars over 0.9 metres in length ranged from 0 to 46 per cent with a mean of 7.7 per cent. About 95 per cent of the cracks exceeding 0.9 metres in length were associated with gross internal cracks, the length of which was closely correlated with the length of the external scar. The proportion of trees only with minor scars (less than 0.9 m in length) ranged from 1 to 22 per cent with an average of 5.5 per cent. Trees without bark scars had no internal cracks, and minor scars were rarely associated with cracks in the wood.

Rot and stain associated with major cracks were not serious. Sawmill conversion of timber from scarred trees indicated that defects were less serious than anticipated and that only in the case of trees with several major scars were the losses extensive. It was estimated that the average loss in value due to stem crack in Grand fir plantations is likely to be less than 1 per cent (Greig, 1969). It thus seems that the occurrence of drought crack will have only a minor effect on the utilisation of Grand fir timber, particularly if affected trees can be removed in earlier thinnings.

Noble fir: drought crack (Plates 13, 15)

Noticeable external bark scarring occurred in 16 (62 per cent) of the 26 stands assessed – a high proportion. Seven of the affected stands had scarring on up to one-third of their trees; nine had scarring on between one-third and two-thirds of their trees. Scarring was rated as slight in 7 stands, moderate in 4 and severe in 5. These ratings indicated more severe scarring than in stands of any other species. Occurrence of scarring appeared to be associated to some extent with increasing crop age and tree size, but not with site and soil drainage nor with rooting depth.

It has been know for some time that external scarring is common in Noble fir plantations but only one detailed study has been made of the relationship between scarring and internal crack. On the one site examined, the length of internal cracks often exceeded that of the associated bark scars. Internal cracks were also found which were not associated with bark scarring. These and other more limited observations suggest that internal cracking is more serious than in Grand fir and that fungal stain and rot are more likely to be associated with cracking. In view of the observed high frequency of occurrence of scarring, it appears that drought crack could well have an important bearing on the utilisation of Noble fir timber.

Sitka spruce: drought crack
Only one stand out of 44 examined in comparisons with adjacent minor species had noticeable bark scarring. Scarring was present on fewer than one-third of the trees and was rated as slight. Thus, severe crack associated with scarring appears to be much less frequent in Sitka spruce stands than in Noble or Grand fir stands. This conclusion must be qualified by results of studies carried out at the Princes Risborough Laboratory (Broughton 1962) and by Day (1954) which show that small internal cracks across one or two rings may occur relatively frequently. Broughton found stem cracks in trees from 12 of the 18 stands he examined. Twenty-two per cent of the total number of trees he examined were affected by small internal cracks but only 5 to 10 per cent of the cracks had given rise to external bark scars.

A study carried out on one site by the Forestry Commission Pathology Section also showed that internal cracks not related to external scars were generally small in comparison with those associated with scarring.

Thus, while there is clearly more drought crack in Sitka spruce than is apparent from the incidence of scars on the bark, the internal cracks are likely to be of far less practical significance than those associated with scarring.

Drought cracks in Other Species
Very little scarring was reported on Norway spruce or Douglas fir and none on Corsican pine nor on any of the larches.

Effect of Drought Crack on Timber Revenues
Losses in Discounted Revenue in stands fairly seriously affected by drought crack are valued at £25, £12, £5 and £5 per hectare for Noble fir, Grand fir, Western hemlock and Sitka spruce respectively, using as a basis the observed reduction in out-turn value from the study on Grand fir.

3.32 Forked stems and Multiple Leaders
Hemlock and Western red Cedar
Forking was observed in roughly 40 per cent of all hemlock and 60 per cent of all Red cedar plots and was more severe in the latter. Field notes show that in a minority of these plots, there were still many trees with multiple forking; in the others, forks occurring early in the life of individual trees had been singled and the scars were successfully callousing over. Forking increased with increasing spacing, with increasing exposure and with diminishing vigour.

If it is assumed that forked trees at first thinning stage are worth 25 per cent less than unforked trees of similar height, the loss in stands with 10 per cent of forked trees is of the order of £5–£10 per hectare; many of the forked stands in Table 11 are likely to be in this category, but in those where forking is more severe, the loss of value is likely to increase more than in proportion to the increased forking.

Most of the forks in Western hemlock occur in the bottom 1.5 metres of the stem and can practicably be singled if required. Red cedar is more likely to fork above breast height than to divide at the base and is therefore less amenable to singling.

Other Species
Forking was rarely encountered in plots of other species.

3.33 Poor Stem Form, Fluting and Buttressing
Defects included as 'poor stem form' include butt sweep, wavy stems and stems showing spiral grain. Table 11 shows that there was a higher incidence of stem defects in hemlock and Noble fir stands than in Grand fir, Red cedar or spruces. However, poor form (basal bowing and crooked stems) was much more prevalent in larch crops than in crops of any other species. Poor stem form in hemlock stands was almost always described as crooked or wavy stems; these were more pronounced than in stands of any other species except the larches. Noble fir has as many stands with butt sweep as with crooked stems but neither of these defects were at all widespread in any stand.

Fluting was most pronounced on hemlock, being present in nearly half the stands visited. It was also quite common in Red cedar stands but buttressing was virtually restricted to Red cedar. For this species, both defects were clearly associated with girth; the bigger the tree the more fluting and buttressing. For hemlock, fluting appeared to increase with increasing altitude of the site rather than with increasing girth of trees. The fluting observed on Noble fir also appeared to be a response to increasing exposure (Plate 3).

The observations in Table 11 were not sufficiently

precise to derive any figures for proportions of logs in different quality grades as a result of poor stem form. However, a slightly lower price might be expected in those hemlock stands with a high proportion of wavy or heavily fluted stems.

3.34 Crown Blast
Symptoms of die-back or of blasting by strong winds were absent in the large majority of stands examined. However, Table 11 shows that about one in ten of the Grand fir, Douglas fir and hemlock plots showed signs of blast, Grand fir being more affected than the other species.

3.4 Crop Stability
3.41 Incidence of Windthrow in Surveyed Stands
Windblow had occurred in between 25 and 60 per cent of the plots of most species except larch and Corsican pine (Table 11, page 48). These latter had fewer plots with blown trees but are likely to have been planted on the drier or better-drained sites. Western hemlock, Red cedar and Norway spruce were less affected than Sitka spruce, Douglas fir, Grand fir and Noble fir. However, in the very large majority of stands, only isolated individual trees had blown down, often associated with local wet soil conditions. Incidence of blow in Noble fir could be related to altitude but no relationship with any site factor emerged for the other species.

3.42 Resistance to Overthrow by Tree Pulling
Appendix III, page 101, gives details of studies in 6 forests where major and minor species were pulled over. While some significant differences were observed on various sites, there was no consistent pattern from which it could be deduced that one species was more resistant to windthrow than another.

Chapter 4
YIELD AND SEED SOURCE:
THE NATURAL RANGES OF THE MINOR SPECIES

Introduction

The growth data obtained in this study have been derived from plantations established when little was known about the effect of provenance on yield and there was no experience of using seed from selected seed sources or seed orchards. Consequently, when considering the future use of any one species in relation to others, the effect has to be estimated of the use of seed of more productive provenance or of seed orchard origin. Table 12 sets out the seed origins of six American and one European species as far as these are known. No details are given for Corsican pine or Japanese larch because for all practical purposes these can be assumed to originate from a homogeneous seed source. Notes on the distribution of the NW American species, relating to Table 12 follow in Section 4.2.

4.1 Potential Improvement

Table 13 summarises the present position in provenance research and the tree improvement programme and also indicates the order of increase in yield which can be expected from the application of the best knowledge and techniques.

TABLE 12

ORIGINS AND QUANTITIES OF SEED OF TWO MAJOR AND FOUR MINOR SPECIES USED IN FORESTRY COMMISSION PLANTATIONS 1919–50

kg

Species	Home Collected	BRITISH COLUMBIA			USA				Total
		Coast	Interior	Not Specified	Washington	Oregon	Other	Not Specified	
Western hemlock	*	125	1	250	14	—	—	14	404
Red cedar	*	71	1	80	18	—	5	15	190
Grand fir	*	282	—	84	221	81	1	951	1620
Noble fir	1306	—	—	—	10	—	—	32	1348

Species	Home Collected	BRITISH COLUMBIA					USA				Total
		QCI	Vancouver Island	N. Coast	Interior	Not specified	Alaska	Washington	Oregon	Not specified	
Sitka spruce	*	24 300	2	11	—	934	1	2 654	2	761	28 665
Douglas fir	*	L. Fraser River					Idaho/ Montana				
		4 482	—	1	24	794	1	5 420	842	2 474	14 038

Species	Home	France	Germany	Italy (Tyrol)	Switzerland	Austria	Romania	Denmark	USA	Total
Norway spruce	7 987**	1 178	3 452	430	2 127	3 239	58	136	4	18 611

Notes. *Not exceeding 100 kg in all. Exact figures not available.
**Almost all in 1942 and 1943.
QCI = Queen Charlotte Islands.

TABLE 13

PROVENANCE AND TREE IMPROVEMENT

SUMMARY OF EVIDENCE AND CONSEQUENCES FOR FUTURE PREDICTIONS

Species	Provenance Present Position	Tree Improvement Programme					Probable Increases in Yield Class due to Improved Seed Source in Plantations to be Established in Next Two Decades	
		Status* of Seed Orchard Programme	Area of Seed Orchard hectares	Area of FC Cat. 'A' managed Seed Stands hectares	Number of selected Plus Trees	Remarks	Increase in Yield Class	Source of Improvement
Western hemlock	Earliest experiments planted in 1961–3 do not indicate any simple pattern of vigour in relation to origin. Washington provenances generally slightly more vigorous than those from British Columbia but differences small. Alaskan provenance slower but better survival.	0	Nil	2	38	Limited vegetative propagation carried out; no extensive breeding programme envisaged.	¼	Selected seed stands
Red cedar	No clear trends from experiments planted in early sixties. No other data.	(1–2)	2	7	59	A small area of combined tree bank/seed orchard available for further plus tree selections.	¼	Selected seed stands
Grand fir	No useful experimental data. From distribution of the species, best growth expected from provenances in low foothills of Cascade Mountains, Washington and Vancouver Island.	0	Nil	2	26	Limited vegetative propagation carried out – no extensive breeding programme envisaged.	¼	Selected seed stands
Noble fir	No data	0	Nil	1	33	As for Grand fir	¼	Selected seed stands
Sitka spruce	Washington provenances give higher yields than Q. Charlotte I. on more favourable sites. QCI plants preferred on frostier and colder upland sites (Lines, 1970)	0	Nil	19	838	Plus tree selection programme and formation of tree banks – well advanced	1	Washington (Oregon) provenance but only on better sites.
Norway spruce	Growth of provenances from SE Europe has consistently been more vigorous than those from C. Europe. Scandinavian provenances grow much less vigorously still.	0	Nil	13	68	Limited selection of plus trees 1948–60. No other breeding work carried out.	½	S E European provenances.

TABLE 13

PROVENANCE AND TREE IMPROVEMENT

SUMMARY OF EVIDENCE AND CONSEQUENCES FOR FUTURE PREDICTIONS

Species	Provenance Present Position	Tree Improvement Programme					Probable Increases in Yield Class due to Improved Seed Source in Plantations to be Established in Next Two Decades	
		Status* of Seed Orchard Programme	Area of Seed Orchard hectares	Area of FC Cat. 'A' managed Seed Stands hectares	Number of selected Plus Trees	Remarks	Increase in Yield Class	Source of Improvement
Douglas fir	Best growth and frost resistance in provenances from high rainfall areas of west coast of Washington (Coast Mts.). Other high rainfall sources in low foothills of Cascades in Washington and Vancouver Island only little less vigorous.	1–2	6	19	506	Seed orchard programme mainly in early stages. Small cone crops available next decade.	½	Selected stands and (later) seed orchards.
Corsican pine	No data other than comparing Corsican pine with other forms of *Pinus nigra*.	0	Nil	104	169	Seed crops will be from selected stands; area to be increased.	¼	Selected seed stands
Scots pine	The provenance collected from the best stands in the locality has so often done well as to be the best general recommendation.	2	16	119	920	All Scots pine seed used by FC will be from seed orchards. The area of stands is likely to be reduced.	1	Seed orchards
Hybrid larch	No relevant data.	2	12	29	26	Seed orchard programme developed for production of F1 seed. Flowering has been disappointing with problems co-ordinating male and female flowers opening.	2	Hybrid larch from seed orchards.

Notes: * Seed orchard programme
      2 Advanced-Trees planted 10 or more years ago – seed likely to be forthcoming in the next decade.
      1 Early stages
      0 None started

Neither provenance research nor the tree improvement programme have paid much attention to the minor species and little improvement as a result can be anticipated. On the other hand, the vigour of future plantations of certain major species, especially Scots pine and Hybrid larch are likely to be materially enhanced.

4.2 Brief Notes on the Natural Distribution of the Minor Species, and Sitka spruce and Douglas fir in Western North America

The natural range of a species indicates how far it has been able to spread by natural regeneration, in the context of the ecology and geographical history of an area. It is no certain guide to the relative performance of often arbitrarily selected provenances of species when established and tended in plantations. Nevertheless, it should be more a cause for surprise if ecological behaviour observable in natural forests is contradicted by performance in plantations than if the patterns agree. See also Wood (1955), Krajina (1965).

Western hemlock, Grand fir and Western red cedar, and also Douglas fir occur associated together over extensive areas in the Coast and Cascade mountains of Washington and Oregon and in the southern-most part of the coast mountains of British Columbia.

The ranges of Sitka spruce and Noble fir are almost mutually exclusive, only overlapping in the foothills of the Cascade Mountains in Washington. Each species within its own range, however, overlaps into the range of Western hemlock, Western red cedar, Grand fir and Douglas fir.

The most marked differences in ranges are firstly in their north-western limits and, secondly, in the easterly and altitudinal limits. (Figs. 19 and 20.)

Northern Limits

In the coastal region, Sitka spruce and Western hemlock extend the farthest north (latitude 61°N), Sitka spruce having extended westwards in Alaska further than Western hemlock. The northern and north-eastern limit of both species is set by a mountain barrier.

Western red cedar's northern limit at 57°N is roughly halfway between the northern limit of Western hemlock and that of Douglas fir. It coincides with the mean summer isotherm (May–September) of 52°F (Anderson, 1952). Douglas fir's northern coastal limit is more difficult to define. In the high rainfall coastal regions it occurs at 51°N but the tree occurs near sea level in sheltered and warm intermediate rainfall sites in the floors and lower slopes of the deep fiord-like valleys in the coast mountains at latitude up to 53°N. In the interior of British Columbia, under conditions with no true counterpart in Britain, the tree extends as far as 55°N.

Grand fir has the most southerly distribution of the six NW American species under review. Its most northerly occurrence is in British Columbia on Vancouver Island and the opposite mainland coast at 50°N. In the interior of British Columbia, the species extends slightly further north to latitude 51°.

Noble fir's most northerly stands are at latitude 48°N. However, it is not a lowland tree and its northerly limit cannot validly be compared with the sea level limits listed for the other species.

Eastern Limits

Sitka spruce is by far the most restricted in the eastwards extension of its native range. Mostly, it is confined to sites of high humidity or moisture status, usually associated with coastal climatic conditions. It is very rarely found further than 160 km inland; in Alaska and British Columbia, commercial stands do not occur much above 600 m. In Washington and Oregon the range of the species narrows southwards, being progressively more restricted to the valley floors of coastal rivers. (Anon, 1965).

In the northern coast region of British Columbia and in South-West Alaska, the eastern limit of Western hemlock and Western red cedar (and Sitka spruce also) is set by the precipitous slopes of the Coast mountains. Both are absent from the plateau land in the rainshadow of the Coast mountains but both reappear along with Douglas fir inland in southern British Columbia in the interior wet belt in the foothills of the Canadian Rocky Mountains. Both hemlock and Red cedar also extend into Northern Idaho.

Douglas fir, although absent from the northern coast of British Columbia, extends much more widely than hemlock and Red cedar on to drier sites in the interior. The climate in such sites is continental in pattern, with warmer summers and colder winters than are experienced in Britain, Grand fir also lies within the range of Douglas fir, occurring in higher rainfall areas rather than on the driest Douglas fir sites. While both species are major constituents of forests in southern interior British Columbia and in Idaho, these are growing under continental, montane conditions with no counterpart in Britain. In the more westerly, coastal parts of its range such as in SE Vancouver Island and in the Puget Sound, Washington, Grand fir grows in a 500–1000 mm rainfall where there is often a marked summer drought, but within such areas, it is restricted to where there is a high ground water table. It is not abundant in the belt of extremely high rainfall along the coast, within the north-western part of its range. Grand fir is a lowland tree, found predominantly below 300 m and rarely above 450 m.

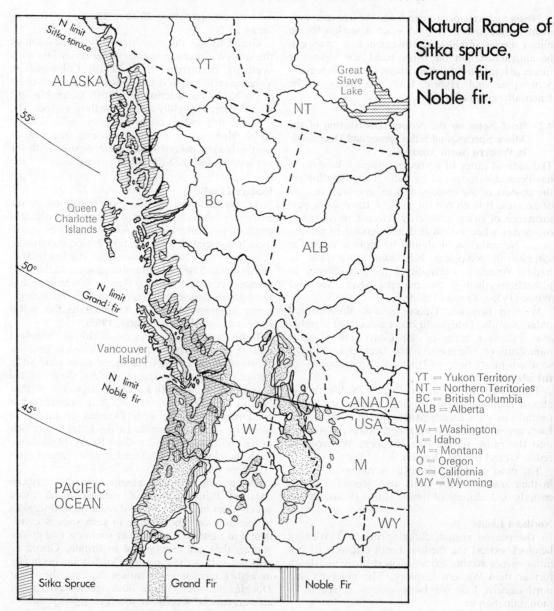

Figure 19. Natural Range of Sitka spruce, Grand fir, Noble fir in Western North America.

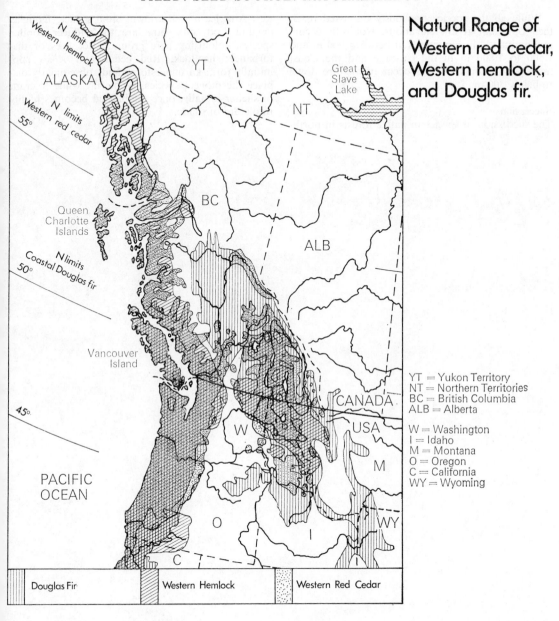

Figure 20. Natural Range of Western red cedar, Western hemlock and Douglas fir in Western North America.

While all the species previously discussed reach their optimum at low elevations, Noble fir occurs predominantly in mountains at between 900 m and 1 500 to 1 800 m in the Cascade and the Coast mountains, though it does occur in mixture down to below 300 m.

Succession

The succession of species in native forests in north-west America is well known. Sitka spruce, Douglas fir and Lodgepole pine are the pioneer conifer species, colonising bare ground after fires or disturbance; hemlock, Red cedar and *Abies* spp. initially form an understorey and subsequently displace the pioneer species. Noble fir, while tolerant of shelter initially, once established becomes highly intolerant of shade.

Chapter 5
DAMAGE BY PESTS AND DISEASES

Introduction
The evaluation of potential economic loss due to pests and disease is more hazardous even than the predictions made in preceding chapters because of the uncertainties surrounding the severity of future attacks. For example, if all hemlock stands are to sustain as severe losses from *Fomes annosus* (*Fomes* butt rot) as in the worst known examples, the prospect for the species is extremely poor. On the other hand, if loss of timber due to butt rot remains at its present level, hemlock is put at only slight disadvantage (Section 5.1). Similarly, for Grand fir on drier sites in the eastern half of Britain, degrade of timber due to formation of 'rotholz' – or a form of compression wood – following infestation by *Adelges piceae* has occurred in a number of stands. Degrade from this cause by itself or in conjunction with factors such as drought crack might be widespread and serious; equally it could be of trivial significance (Section 5.41).

5.1 Fomes Butt Rot, *Fomes annosus*
Fomes butt rot is a potential threat to many conifer species in British forestry, a threat which varies according to the relative susceptibility of conifers and to certain site characteristics which are associated with serious losses. *Fomes* butt rot is not and is unlikely to become a problem of vast magnitude in all forests but it is important in current and future crops in some localities. If a resistant species is found which can be grown on high-risk sites with minimum loss by decay, then the losses due to the disease can be kept under strict control in British forests.

Area at Risk from *Fomes*
Observations between 1930 and 1960 for the incidence of *Fomes* butt rot (Peace, 1938; Low and Gladman, 1961) strongly suggested that the disease was a serious problem in plantations with a previous history of conifer crops. Current investigations reported in Appendix II support this view. The area of Forestry Commission plantations growing on ground with a previous conifer history total is estimated at about 24 000 ha; to this must be added an area of about 26 000 ha, being that small proportion of older first rotation crops likely to have been seriously infected by *Fomes* before the introduction of protective stump treatment. However, the losses by decay in this category of crop are expected to be much less than those on second rotation areas.

Serious losses from *Fomes* are also expected to occur, not in the present crop but in the future crops on areas where pine is to be replaced by more productive species susceptible to *Fomes* butt rot. The area involved is difficult to forecast at this time but it is unlikely to be less than 40 000 ha.

The area at risk in the three categories together totals some 90 000 ha, an area which is relatively small compared to the Forestry Commission's total acreage. Similar areas in private estates could bring the total to between 160 and 200 000 ha. However, on much of this land, both Grand fir and Western hemlock could grow rapidly.

Recent Investigations
In about 200 of the older stands assessed for growth as part of the survey of minor species, timber cores were taken by Pressler borer and were subsequently examined for *Fomes*. Very little infection was, in fact, detected, and no conclusions could, therefore, be reached concerning either the differential resistance of species or the effect of site factors on infection.

More detailed studies of losses of timber in eleven plots of Western hemlock and Grand fir are described in Appendix II. Results show that losses in utilizable timber volume in 30 to 45-year-old Western hemlock stands can amount to 20 per cent of standing volume. In contrast, losses in Grand fir stands of similar age were negligible. Although the evidence presented in Appendix II is limited, it consistently indicates that Grand fir when growing on *Fomes*-infected sites is able to resist butt rot infection, while hemlock can suffer substantial loss. More recent studies (Burdekin 1970) have shown that *Fomes* can invade trees in Grand fir stands and that, while the number of such stands so far discovered is very small, the amount of decay can be substantial.

In a recent sawing study, 23 out of 30 Grand fir logs (10 from each of 3 forests) contained some bacterial wet wood. Its extent was not determined. The significance of bacterial wet wood is not fully understood but it is possibly linked with the resistance of Grand fir to *Fomes* butt rot.

No comparable evidence is at present available to assess the relative susceptibility of other conifer species. However, a subjective ranking of the relative susceptibility of conifer species to *Fomes* butt rot is:

Western hemlock ⎫
Sitka spruce ⎪
Western red cedar ⎬ relatively susceptible
Norway spruce ⎪
Larch ⎭

Noble fir ⎫
Douglas fir ⎬ relatively resistant
Grand fir ⎭

Pines   resistant

An evaluation of the economic losses caused by *Fomes* butt rot has recently been made, based on a number of subjective estimates together with the best information currently available. The loss in Discounted Revenue in future crops on the 24 000 ha of land with a previous history of conifer and on the 40 000 acres where pine will be replaced by species relatively susceptible to butt rot was estimated to be 10 to 15 per cent; obviously the losses in some plantations could be negligible and in others much greater than this.

5.2 Honey fungus (*Armillaria mellea*)
This fungus kills recently established trees, particularly where conifers have been planted in old hardwood areas. The damage, while conspicuous, is very rarely sufficient to necessitate additional beating up, nor does it often kill trees in vigorous plantations once into the thicket stage. Loss of timber through rot is of minor importance as it seldom travels more than two feet or so up the stem. Peace (1962) drew up a list of relative susceptibility to killing and decay. However, relative susceptibility to Honey fungus is rarely likely to influence the choice of species decisively.

5.3 Other Diseases Affecting Choice of Species
Western red cedar is subject to attack in the nursery by *Didymascella thujina* but this can be successfully controlled by the use of a fungicide which adds only a little to the costs of production. Western hemlock and Grand fir have no stem or needle disease known to cause serious damage. Noble fir can develop darkly stained heart wood but this is not clearly associated with any specific fungus.

Some of the major species are, in a few localities, seriously affected by certain diseases; such susceptibility could influence subsequent choice of species. For example, *Peridermium pini* can cause losses in Scots pine in Nort-East Scotland and *Brunchorstia pinea* can cause serious damage to Corsican pine at its climatic limit in North-East England and elsewhere. However, on very few of such sites are any of the minor species viable alternatives.

In another recent study, it was noted that small dead knots in boards from 20 to 30-year-old Western red cedar were commonly rotten, leading to downgrading of the boards in spite of the timber around the knots being sound and working well. This suggests that early pruning might enhance the value of sawlog butts of Western red cedar.

5.4 Insect Damage
While many insects have as their host plants the common or minor species of conifers, very few are both sufficiently numerous to be serious potential pests and sufficiently selective in their feeding, to influence the choice of species. The Pine looper, *Bupalus piniarius*, is possibly the most important of these; the choice of species to replace Scots pine could be influenced in favour of alternative species in areas where the pine has been subject to repeated heavy defoliation. Repeated defoliation of Sitka spruce by *Elatobium* (*Neomyzaphis*) *abietinum* and any subsequential loss of increment might also be sufficient to turn the balance in favour of other species. Both these possibilities, however, are of small significance at present.

The one insect which is associated with the premature death or die-back of any of the minor species is *Adelges piceae*, which attacks Grand fir.

5.41 Damage to Grand fir and Noble fir by *Adelges*
Adelges piceae may inhabit the bark of main stems of many species of *Abies* and can cause abnormal wood formation, similar to compression wood and widely referred to by the German name 'rotholz' (i.e. red wood). Infestation can also occur on twigs of certain North American firs when it may cause gouty swellings. The insect which has damaged Grand fir is thought to be a different strain from the strain of *A. piceae* causing damage to *Abies alba* (Busby 1962, 1964).

Resin bleeding on the stem is a typical symptom of heavy attack by *Adelges piceae* but is not sufficient symptom by itself, as some diseases can also cause resin bleeding. Severe stem infestation leads to the formation of 'rotholz' and this in turn restricts water movement to the upper crown of the trees. As a consequence, trees may die or the crowns die back (Doerksen & Mitchell, 1965). *Adelges* infection may be greater if trees are in exposed situations or are subject to acute drought.

Grand fir crops have failed at several places after reaching 15 m or more in height. In the Bedgebury Forest Plots, many trees in the Grand fir plot, planted in 1931, have lost most of their foliage, defoliation having been initiated by severe summer drought. The trees several years later were found to be heavily infested with *A. piceae*, the whole plot dying back and failing over a period of about ten years. Failure has also been reported from Redleaf Estate, Kent; Craibstone, Aberdeen; Beddgelert, Caernarvon, and Winderwath, Cumberland. In all of these instances, one or more of the factors of drought, exposure and the presence of *A. piceae* are believed to have brought about the failure.

A survey in Grand and Noble fir stands in England; Scotland and Wales in 1965 showed *A. piceae* to be

DAMAGE BY PESTS AND DISEASES

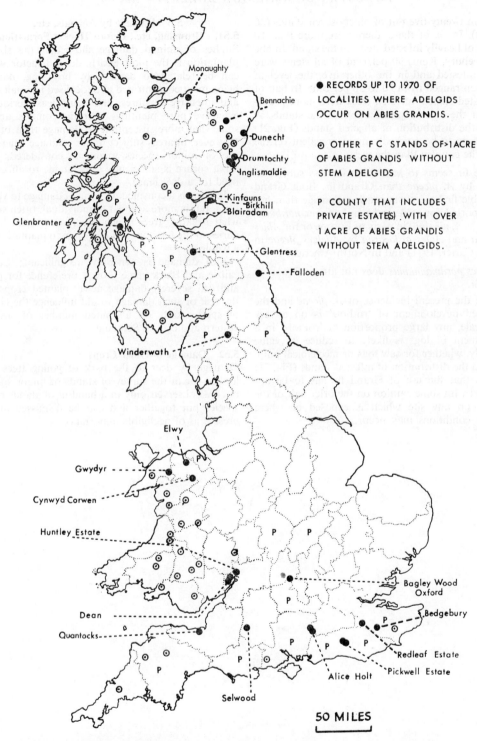

Figure 21. Map of the distribution of Grand fir stands infested by the insect *Adelges piceae*, throughout Great Britain.

present in twenty-five out of ninety-seven stands (26 per cent). In six of these, there were more than 10 per cent of heavily infested stems in the stand. In one (at Bedgebury, Kent) 90 per cent of all stems were heavily infested and in the other five, the level of infestation ranged from 20 to 50 per cent. In half of the infested stands, affected trees were more prevalent near the more exposed edges of the stands. A map of the distribution of affected stands (Fig. 21) shows these to be markedly more prevalent on the east of the country than the west.

Noble fir seems to have been far less subject to damage by *A. piceae* than Grand fir. Both Grand and Noble fir have sustained less damage than has been reported for *Abies alba* by *Adelges nordmannianae* (= *A. nüsslini*) in parts of Europe or for *Abies balsamea* and *Abies lasiocarpa* by *Adelges piceae* in Britain (Carter, 1971) and in North America.

Adelges nordmannianae does not attack Grand or Noble fir.

While the present incidence of *A. piceae* and the associated development of 'rotholz' is on a very small scale, any large proportion of 'rotholz' in a consignment of logs is likely to reduce its value markedly, whether for saw logs or mechanical pulp. This and the distribution of infested stands (Fig. 21) suggests that the use of Grand fir has to be approached with some caution on the drier half of the country on any site which is exposed or where drought conditions may occur.

5.5 Damage by Animals, etc.

5.51. Browsing, etc., Before Thicket Formation

Studies of animal damage show that the size of plantation is the most clearly defined factor which can be related to damage by browsing, damage being most concentrated and localised in small plots and intensive mixtures, and being least serious in extensive pure plantings. Certain conifers appear somewhat more susceptible to damage than others. However, in areas subject to deer damage where the use of minor species might be considered, only Sitka spruce is sufficiently unpalatable to affect the need for a deer fence.

Fraying is a conspicuous form of damage to young trees, but is more a threat to individual, often somewhat isolated, trees which occur on natural boundaries to the territories of deer than to plantations as a whole.

Bird damage is generally insignificant; Capercailzie and Black game show a preference for pines and can severely damage newly planted crops but the risk of such damage would influence the choice of species only on a limited number of sites in eastern and central Scotland.

5.52 Damage in Older Crops

Damage by deer to the bark of young trees was looked for in the survey of stands of minor species but was observed only in a handful of stands of all species put together and can be dismissed at the present as of negligible importance.

Chapter 6
WOOD PROPERTIES, UTILIZATION AND MARKETING

Introduction

In this chapter, it is assumed that new home-grown timbers will in due course find markets in Britain provided that their wood properties are acceptable for their intended use and that sufficient volume is produced to interest timber merchants and wood users.

In general the information available on British-grown wood of these trees is limited. What is known about the wood properties of the four species has recently been summarised by Brazier (1973) and material from this paper has frequently been used in preparing the following sections. The results of recent investigations into the sawmilling and woodworking properties are given by Priest (1973) and Greenwood (1973). The various joint studies by the Princes Risborough Laboratory of the Building Research Establishment (formerly the Forest Products Research Laboratory) and the Forestry Commission have provided valuable data on specific gravity and moisture content of home-grown timber of all four minor species which are the subject of this report, as well as comparable data for Sitka spruce, Norway spruce and Douglas fir. Figures are given in Table 14A, and refer for the most part to timber cut from trees 30–50 years old. Summarised strength data for the same seven species given in Table 14B are taken from Forest Products Research Laboratory Bulletin 50, *The Strength Properties of Timbers* (Lavers, 1969).

TABLE 14A
NOMINAL SPECIFIC GRAVITY AND PERCENTAGE MOISTURE CONTENT OF WOOD FROM STANDS OF MINOR AND MAJOR CONIFER SPECIES GROWN IN BRITAIN
Data from Princes Risborough Laboratory. Based on Published and Unpublished Studies

		Minor Species				Major Species		
		Western hemlock	Western red cedar	Grand fir	Noble fir	Sitka spruce	Norway spruce	Douglas fir
Nominal specific gravity	Overall mean	0·39	0·32	0·31	0·32	0·34	0·33	0·43
	Range of stand means	0·36–0·42	0·31–0·39	0·28–0·34	0·30–0·35	0·31–0·40	0·30–0·35	0·38–0·47
Moisture content %	Overall mean	145	178	179	210	138	166	101
	Range of stand means	90–172	141–228	126–237	182–229	99–160	143–183	86–113

TABLE 14B
STRENGTH DATA FOR WOOD OF MINOR AND MAJOR CONIFER SPECIES GROWN IN BRITAIN
Data from FPRL Bulletin 50, and based on tests of small clear specimens at 12 per cent moisture content.

Species	Sample	Modulus of rupture N/mm^2	Modulus of elasticity N/mm^2	Compression strength N/mm^2	Hardness N	Shear N/mm^2
Western hemlock	15 trees	76	8000	41·3	2580	10·6
Western red cedar	10 trees	65	7000	35·0	2000	8·5
Grand fir	11 trees + 20 joists	57	7000	30·1	1780	7·7
Noble fir	5 trees	63	8100	31·0	2000	9·3
Sitka spruce	54 trees + 50 joists	67	8100	36·1	2140	8·7
Norway spruce	188 trees	66	8500	34·8	2000	9·4
Douglas fir	54 trees	91	10500	48·3	3420	8·9

TABLE 15
RESULTS OF TESTS ON WESTERN HEMLOCK THINNINGS
(Based on Forest Products Research Laboratory Reports on Consignment Nos. 809, 873 and 949)

Source	Age (yrs.)	Mean Diameter cm	Mean Moisture Content %	Mean Nominal Specific Gravity	Mean Density 12% Moisture Content kg/m^3	Mean Growth (rings/cm)
Dunster Estate, Somerset	54	71	111	0·40	480	3·0
Novar Estate, Ross and Cromarty	55	102	116	0·39	464	2·4
Dyfi Forest, Merioneth and Montgomery	29	75	146	0·37	448	1·9

The incidence of defects such as poor stem form and drought crack, which could affect marketing, are discussed in Chapter 3, Section 3.3.

6.1 Western Hemlock
6.11 Weight and Strength Properties of Hemlock

Wood samples from 12 randomly selected young pole-stage stands in the most recent study had a mean nominal specific gravity of 0.39 and green moisture content of 145%. The nominal specific gravity of three consignments of Western hemlock thinnings tested at the Princes Risborough Laboratory between 1955 and 1959 (Table 15) also came within the range of these more recent samples. (Reports on thinnings, Consignment Nos. 809, 873 and 949.)

All these values for hemlock are intermediate between those for home-grown Sitka spruce and Douglas fir also given in Table 14A.

Results of other tests on the three earlier consignments are summarised in Table 15.

Available strength data for home-grown Western hemlock are given in Table 14B. Individual results for the thinnings consignments showed that the timber from Dunster was more or less equivalent in weight and strength to the average values for Western hemlock timber grown in the USA but was not quite as dense or as strong as Canadian grown timber; the timber from Novar was somewhat lighter and weaker than that from North America, while the Dyfi material was lightest of the three lots and relatively brittle (probably due to its being both young and fast grown). Much of the sawn material from the thinnings obtained at Dunster and Novar was graded as suitable for general structural work (Grade II). However in the most recent saw-milling investigation, a high proportion of timber was placed in Grades III and IV, i.e. only suitable for uses (packaging, etc.) not requiring consistent strength properties.

6.12 Sawing, Seasoning and Wood Working Properties

No special problems were encountered when sample logs from six sites in different parts of Britain were sawn using standard commercial machines and sawblades.

Home-grown Western hemlock timber can be seasoned rapidly using PRL Kiln Schedule L (Anon., 1969) with little checking or end splitting, but cup is often a serious defect and twist can be severe. Twist is associated with the presence of spiral grain, and although no observations have been made of the pattern of grain development it appears that spiral growth can be well developed, especially in juvenile wood. A visual appraisal of material used in machining trials suggested that twist is more severe with Western hemlock than with either of the firs or Western red cedar. Shrinkage and drying is moderate and movement of dry timber with changes in humidity is rated as medium. The timber darkens somewhat to a fairly uniform pale brown colour on kiln drying. Imported timber is said to air-season slowly.

In a recent, not yet published study on home-grown timber, a correct cutting angle was prescribed as essential for a good finish. Even so, knots and spiral grain led to tearing. Numerous small knots were inclined to damage cutting edges. Nailing properties were, on the whole, good.

Home-grown timber is likely to be similar in durability to imported timber and to be rated as "non-durable" in relation to fungal attack.

No information is available regarding the permeability to preservatives of home-grown hemlock timber. It seems likely to be similar to that of imported hemlock which, like Sitka spruce, is considered resistant to impregnation.

6.13 Appearance of Wood

Home-grown wood of Western hemlock is generally pale in colour with a somewhat brownish tint (Plate

11). Normally there is no colour contrast across the end section of a log but occasionally the core wood is a little darker than the outer wood and a watermark-like stain may be observed in some timber. Growth rings are not a conspicuous feature of machined surfaces and the uniform character of clear wood appears to be due to the lack of colour contrast between early and late wood tissue. In rapidly grown wood, the annual rings contain a high proportion of early wood, merging gradually into the late wood, giving rise to a fairly uniform texture. In slower grown material, the late wood forms a higher proportion of the ring, but the gradual transition from early to late wood remains. Resin streaks occur occasionally and are a characteristic feature of the timber.

Knots are conspicuous and are often large, but smaller knots also occur, particularly in timber cut from near the centre of the tree. Dead knots are particularly noticeable because of the associated enclosed bark which appears to be a characteristic of the species. A feature of boards cut near the centre of the tree is irregular grain associated with a wandering pith.

6.14 Pulping Properties of Hemlock

Laboratory-scale pulping trials have been carried out at the Princes Risborough Laboratory using material from four trees chosen at random in each of two plantations in S.W. England and Wales (unpublished data). Results for disc refiner pulp showed Western hemlock to be slightly inferior to Sitka spruce in both strength and brightness, virtually equal to Scots pine, slightly better than Douglas fir and much superior to larch. With chemical pulping (both Kraft and two-stage sulphite processes), hemlock gave pulps with physical properties only very slightly inferior to those from Sitka spruce, but with a yield some 4–5 per cent lower.

It was concluded that hemlock and spruce pulp could probably be used in mixture for most purposes, and that it is technically feasible for the two species to be pulped in mixture. Western hemlock is in fact a long-established major source of chemical and mechanical pulp in Western USA. According to Casey (1960) "In the western states, Western hemlock, Sitka spruce and the true firs are the most desirable species. Western hemlock produces an excellent pulp of high strength which can be bleached without difficulty. . . . Sitka spruce is slightly superior to hemlock and fir and is highly prized because of its good colour". A recent review of mechanical pulping (Gavelin, 1966) states that "The best softwood species for groundwood are black and white spruces and Western hemlock. . . . Western hemlock is often bleached due to its dark colour caused by tannin, but it gives pulp of good strength".

6.15 Utilisation and marketing of Hemlock

It appears unlikely that there would be any real problem in utilising home-grown Western hemlock timber if available in quantity, nor is there any foreseeable limit to the amount which could be sold for most purposes, i.e. sawn wood, pitwood, chipboard, fibreboard and chemical pulping. For groundwood pulping for newsprint in Britain, hemlock is not at the present accepted, but see the preceding section. However, little hemlock is yet marketed; if large volumes become available, marketing opportunities will improve (see also Section 6.6).

In the past, the timber trade in Britain has been prejudiced against the use of imported Western hemlock. This was due mainly to two reasons: firstly, it air-seasons slowly and has arrived in this country in a comparatively green condition, and secondly, it is common practice for a proportion of less dense, weaker timber of various western *Abies* species to be included in Western hemlock shipments, the mixture being known commercially as "hembal". Difficulties associated with these practices appear to have been resolved as the timber now has a good name and is used extensively in timber engineering. As far as the lower grades of imported timber are concerned, recent prices for hemlock do not appear to have differed much from those for Sitka spruce but have been considerably lower than those for Douglas fir.

In western North America, hemlock timber is widely used for such purposes as general structural work, flooring and boxmaking.

6.2 Western Red Cedar

6.21 Weight and Strength Properties of Red Cedar

Mean values of 0.32 for the nominal specific gravity and 178 per cent for the green moisture content are given in Table 14 for the wood from young Red cedar stands in Britain. These values are based on samples from nine randomly selected plantations. The relatively low specific gravity is similar to that of home-grown Sitka and Norway spruce and also that quoted for Red cedar grown in Western North America (Anon, 1955); however, it is well below that of home-grown Douglas fir.

In 1955, tests were carried out at the Princes Risborough Laboratory on two consignments of Red cedar thinnings and details of wood properties are summarised in Table 16.

The moisture content figures given in Table 16 are considerably lower than those given in Table 14A, probably as a result of the two stands being older than the stands sampled to provide the data in the earlier table.

TABLE 16
RESULTS OF TIMBER TESTS ON WESTERN RED CEDAR THINNINGS
(Consignment Nos. 786 and 814, 1955)

Source	Age (yrs.)	Mean Diameter cm	Mean Moisture Content %	Mean Nominal Specific Gravity	Mean Density kg/m³	Mean Growth Rate (rings/cm)
Novar Estate, Ross and Cromarty	49	89	114	not known	376	2·8
Alice Holt, Hampshire	48	89	72	,,	400	3·3

Table 14B gives average strength data obtained from the limited tests which have been carried out. Wood from the two thinnings consignments appear to have mechanical properties slightly superior to those of wood imported from Canada.

6.22 Sawing, Seasoning and Wood-Working Properties

There was no difficulty in sawing sample logs from six sites in different parts of Britain, using standard commercial machines and saws. However, slabs from the fluted butt logs were difficult to handle. In the earlier study of thinnings material, excessive taper of logs from Alice Holt Forest led to high conversion losses.

Using PRL Kiln Schedule J, Western red cedar kiln dries well with little degrade apart from some tendency to collapse. Splitting and checking are usually minimal, cupping slight and bowing insignificant. Spring and twist are not usually marked but can be severe on occasional pieces. Both shrinkage on drying and subsequent movement with changes in humidity are small.

Working properties of home-grown timber have been found to be reasonably good, apart from some tearing of the grain in the vicinity of knots and in the knot wood itself.

In the saw-milling investigation, well over half of the sawn timber produced was graded as suitable for general purpose use (Grade III). However, a substantial quantity fell within Grade II (general structural use) and the proportion was distinctly higher than for any of the other three minor species.

Home-grown timber has been rated as "moderately durable", i.e. moderately resistant to fungal attack. In this respect it is slightly poorer than imported material which is rated as "durable".

6.23 Appearance of Wood

Unlike Western hemlock and the silver firs, Western red cedar has a distinctive coloured heartwood. This is typically a light russet brown which contrasts with the white sapwood. The sapwood band is narrow and, as a result, a high proportion of converted timber consists of heartwood even in fairly young trees. Growth rings are not a conspicuous feature because of the heartwood colour. The growth ring structure shows a gradual transition from thin-walled early wood to a very narrow band of late wood. The wood has a fairly high natural lustre and a characteristic odour. (Plate 10).

Knots, although frequent, are mostly of small to medium size and live. They tend to be distributed along the sawn timber rather than grouped in regular whorls. See also Section 5.3, rot of knots.

6.24 Pulping Properties of Red Cedar

There have been no pulping trials of home-grown Red cedar wood. Pulping of the species in Western North America is discussed in detail by Weathern (1959) and Gardner (1963) and the following information is derived from these papers, and from limited experience in British board mills.

Red cedar timber is unsuitable for groundwood pulping because of the dark colour of the heartwood and consequent bleaching problems, and because of debarking difficulties. Kraft pulping gives a good pulp, but yield is low because of the low wood density, and a longer cooking time is required in comparison with Western hemlock. In addition, pulping of cedar results in accelerated corrosion of digesters, etc., due to certain extractives in the wood so that stainless steel equipment is required if cedar is to form a high proportion of the wood used in a mill. To some extent these drawbacks are compensated by the desirable properties of the Red cedar pulp; on average, Kraft mills use 15-30 per cent cedar and at least one American mill is running exclusively on cedar. However, the quantities of timber consumed are of the order of several hundred thousand cubic metres per annum; there is no prospect of being able to set up a Kraft mill of such a capacity in Britain.

Sulphite pulping of cedar is unsatisfactory. The only likely outlet for Red cedar pulpwood is in mixture with other conifers for board production, provided the timber can be debarked easily.

6.25 Utilisation and Marketing of Red Cedar

Because of the colour and other properties of its

wood, Red cedar does not group conveniently with any of the more common conifer species grown in Britain. At present, the relatively small volumes of Red cedar in the small roundwood sizes can be sold for fencing and estate work, but a sizeable increase in production could lead to difficulties in disposal in many parts of the country unless an entirely new outlet is developed. The species is likely to remain unacceptable for Kraft pulping even in mixture with other species, since nowhere in Britain is there sufficient wood to justify setting up a Kraft mill. Red cedar's colour and corrosion risk make it undesirable for use by the fibre and particle board industries; in addition, the fibrous nature of Red cedar bark has led to difficulties in bark removal by rotary peelers.

In view of the relatively high prices currently paid for imported sawn timber, there is probably no limit to the quantity of good quality saw logs which could be marketed, but production of saw logs would hardly be economic if there were no satisfactory outlets for smaller sizes of roundwood. In general, it appears that any sizeable increase in the scale of planting of Western red cedar would lead to considerable marketing problems in the future.

Imported Red cedar has a high reputation for exterior use in house building and roofing, mainly because of its easy working properties and natural durability when not in contact with the ground. However, the best grades which form the bulk of the cedar exported from Western North America are selected from the outer heartwood of trees often 100 years old; much of what remains is of low grade and difficult to sell.

6.3 Grand fir
6.31 Wood properties of Grand fir

Table 14A gives mean values of 0.31 for the nominal specific gravity and 179 per cent for the green moisture content of wood from young plantations in Britain. These data are based on samples from eight randomly-selected stands in various parts of the country. Young Grand fir timber is on average almost as dense as comparable material from plantations of Sitka spruce and Norway spruce, and markedly less dense than Douglas fir. It has a distinctly higher moisture content than any of these. It is probable that the low specific gravity is associated, at least in part, with the very rapid radial growth found in most of the stands sampled.

Between 1955 and 1959, three consignments of thinnings were examined at the Princes Risborough Laboratory and the results are summarised below in Table 17.

Two of the stands concerned were older than those sampled to provide the data given in Table 14A. In both, specific gravity values were within the range given in Table 14A. However, the moisture content of wood from these older stands was much lower than that found in the recent investigation.

6.32 Sawing, Seasoning and Wood-working Properties

No special difficulties were encountered when standard commercial equipment was used to saw sample logs from six sites in different parts of Britain. Where severe drought cracks were present, slabs and boards sometimes split on the saw causing handling problems and slightly lower production.

Until recently it was thought that the occurrence of drought crack might have a serious effect on the conversion of home-grown Grand fir sawlogs. These fears have been very largely discounted by the detailed investigation, described in Section 3.31, which indicated that average conversion losses due to this cause were likely to be very small (Greig, 1969). During sawmilling connected with that investigation the timber sawed well, and in the view of the sawyer was better in sawing and finish than Sitka spruce, and comparable with Norway spruce.

Using PRL Kiln Schedule L, home-grown Grand fir timber has been dried quickly and well, although with some tendency to collapse. Only very slight checking and end splitting occurred, while spring and bow tended to vary between consignments, but were generally slight. Cupping can be marked, particularly in material cut from young trees, and the degree of twisting was variable. Shrinkage and drying

TABLE 17
RESULTS OF TESTS ON THINNINGS OF GRAND FIR
(Consignments 830, 835 and 948; 1955–59)

Source	Age (yrs.)	Mean Diameter cm	Mean Moisture Content %	Mean Nominal Specific Gravity	Mean Density kg/m^3	Mean Growth Rate (rings/cm)
Novar, Estate, Ross and Cromarty	54	157	89	0·34	408	1·9
Dunster Estate, Somerset	41	97	101	0·34	400	2·1
Dyfi Forest, Merioneth and Montgomery	30	Not known	198	0·31	368	1·7

was moderate, but movement with changes in humidity were rated as small.

In wood-working trials, home-grown Grand fir timber planed reasonably well, particularly with a reduced sharpness angle on the cutters. Tearing was liable to occur where knots and spiral grain were present. Knots also tended to damage cutter blades. The results of machine moulding were unsatisfactory.

When sawn material from the thinnings consignments were studied, the timber was thought to be suitable for all purposes for which home-grown Sitka spruce might be used. However, it is doubtful if this conclusion is still valid in the light of present day thinking. In the more recent saw milling investigation, a very high proportion of the sawn timber was placed in Grades III and IV, i.e. suitable for general purpose use where strength is not important.

Both home-grown and imported Grand fir timber is rated as "non-durable". Resistance to preservative impregnation appears to be lower than for Sitka spruce timber.

6.33 Appearance of Wood

Home-grown wood of Grand fir is typically pale and almost white when dry, lacking the lustre of freshly machined Sitka spruce wood. The heartwood tends to be pale purplish-brown in colour in fresh wood, but fades on exposure and drying to become barely noticeable in dry wood. The core wood may have a greyish colouration possibly associated with bacterial attack. Despite its often rapid growth, the timber is fairly firm and uniform in texture. Like the spruces, Grand fir has a pattern of growth which results in a broad zone of early wood merging gradually into late wood which is usually confined to a narrow band. (Plate 8).

Knots are a conspicuous feature, often whorled but with some smaller internodal knots also present. Cores of darker wood, particularly in the larger knots, often make knots look smaller than they are. Compression wood is an occasional feature, appearing on planed surfaces as pale brown bands. Longitudinal splits due to drought cracking are occasionally present, and may show slight dark staining.

6.34 Pulping Properties of Grand fir

In Britain, small scale pulping tests have been made on wood from two samples of four trees from plantations in SW England and in Wales (unpublished data). Disc refining of Grand fir gave a pulp only slightly inferior to that for Sitka spruce and better than that for Western hemlock. With the two-stage sulphite process, pulp yield and strength for Grand fir were slightly lower than for either Western hemlock from the same forests, or for Sitka spruce of equivalent density.

In general, it appears that substitution of Grand fir for the same roundwood volume of Sitka spruce in mechanical or chemical pulping is likely to result in some reduction in yield and strength of pulp produced; for many purposes however, the strength differences may not be of commercial significance. Chemical and mechanical pulping of the species in mixture is technically feasible.

There appears to be little information available on the pulping specifically of Grand fir in North America, although its properties seem likely to be similar to those of *Abies balsamea* which is widely used for both chemical and groundwood pulps.

6.35 Utilisation and Marketing of Grand fir

It does not appear that there would be any real problem in utilising large quantities of home-grown Grand fir timber in the smaller sizes and no limit to the market is foreseen. As a "whitewood", it groups well with the spruces for fibre and particle board production, and chemical pulping. The density of Grand fir wood supplied for groundwood pulping differs little from that of the spruce normally used and should therefore be acceptable in mixture with Sitka and Norway spruce.

The price for the very limited amount of Grand fir currently being offered for groundwood pulp in Britain appears to be about $3\frac{1}{2}$ per cent lower than for other species.

Sawn timber of Grand fir would appear to be suitable for use in place of lower grades of imported "whitewood", where high strength is not essential. However, if the saw-miller has to take the risk of drought crack affecting conversion, he may offer the grower a somewhat lower price on this account (see also Section 3.11).

In Western North America, Grand fir timber is regarded as a general purpose timber for use where strength is not of importance. (Anon, 1955).

6.4 Noble fir

6.41 Weight and Strength Properties

Recent tests on wood samples from eight randomly selected stands gave mean values of 0.32 for the nominal specific gravity and 210 per cent for the green moisture content (see Table 14A). Noble fir wood is thus similar in density to that of Grand fir and Western red cedar, and considerably less dense than comparable material of Sitka spruce and Norway spruce. The very high moisture content is well above any of the values recorded for other species.

Only one consignment of Noble fir thinnings has been examined at the Forest Products Research Laboratory; details are given in Table 18.

Strength data from a very small number of tests (see Table 14B) suggest that Noble fir timber is distinctly stronger than that of Grand fir and only marginally weaker than timber of Sitka spruce.

TABLE 18
RESULTS OF TESTS ON NOBLE FIR THINNINGS
(Consignment 950, 1959)

Source	Age (yrs.)	Mean Diameter cm	Mean Moisture Content %	Mean Nominal Specific Gravity	Mean Density kg/m³	Mean Growth Rate (rings/cm)
Dyfi Forest, Merioneth and Montgomery	30	Not known	186	0·34	400	2·0

6.42 Sawing, Seasoning and Wood-working Properties

In a saw-milling trial, sound Noble fir logs from six sites in different parts of Britain were sawn without difficulty using standard commercial equipment. Logs with severe drought cracks present produced weakened slabs and boards which caused handling problems and lowered production.

The occurence of drought crack in Noble fir has been discussed in Section 3.31; from the available evidence, crack is likely to be more serious than in Grand fir. There is thus the risk of appreciable volume and value losses during the conversion of Noble fir for sawlogs.

Conversion is also likely to be affected by the higher rate of taper of Noble fir logs in comparison with most other species.

Only one consignment of Noble fir timber was kiln dried at the Princes Risborough Laboratory. Its behaviour was reasonably satisfactory, using PRL Kiln Schedule L, and was similar to that of Grand fir timber apart from a greater tendency to bowing. No data are available on the shrinkage and movement of Noble fir timber.

In wood-working trials, home-grown Noble fir timber behaved similarly to Grand fir timber (see Section 6.32).

When sawn material from the thinnings consignment was studied, it was tentatively assessed as being suitable for joinery purposes but not for constructional work. In the more recent saw-milling study, a very high proportion of the sawn timber fell in the general purpose Grades III and IV, as did most of the Grand fir timber.

Noble fir timber (both home-grown and imported) is rated as "non-durable". Like Grand fir timber, it appears to be less resistant to preservative impregnation than are spruce and hemlock.

6.43 Appearance of Wood

There is little difference in appearance between wood of home-grown Noble fir and that of Grand fir (described in Section 6.33), apart from the latter having slightly more conspicuous late wood zones. Splits due to drought cracking occur more frequently than in Grand fir and are more likely to be associated with fungal stain and rot (see Section 3.31).

6.44 Pulping Properties of Noble fir

Little information could be found on the pulping properties of Noble fir, although it is known to be used to some extent for pulping in North America. No tests have been carried out on home-grown material. In view of its similarity to Grand fir, it is probable that results of pulping by the various processes would also be similar, i.e. pulping satisfactory, but yield and strength properties somewhat lower than those for Sitka spruce.

6.45 Utilisation and Marketing of Noble fir

As a "whitewood", Noble fir can be grouped with the spruces for marketing purposes in most of the normal small roundwood markets. However, it is not normally accepted for groundwood pulping – apparently due to trouble in the past with stained wood. Sawn timber would probably be suitable for use in place of the lower grades of imported "whitewood" but conversion losses due to internal cracking might make it unpopular with timber merchants. Bearing in mind these possible drawbacks, the market could probably absorb all Noble fir wood offered for sale, but at a lower price than for most other species.

In Western North America, Noble fir timber is regarded as being of relatively high grade and is used in quantity for a range of purposes including joinery and general constructional work (Anon, 1955).

6.46 Noble fir Foliage

While no valuation has been made in Britain, the foliage of Noble fir has in recent years commanded very high prices in Denmark and Germany. The demand has been of sufficient duration for experiments to have been undertaken in Denmark to determine the effects of added fertilizer and weed killers on the yield of foliage. Holstener-Jørgensen (1973) reports yields of 5 tonnes per hectare for 30–40 year old crops. Figures for total crown weight from the Noble fir wind stability trial described in Appendix 3 suggest that for younger crops, 3 tonnes per ha might be more appropriate.

The significance of this market in relation to plantations in Britain is an open question.

6.5 Factors Affecting Timber Prices in Britain
6.51 Home-grown Log Prices

Factors affecting the selling prices of sawlogs at stump currently seem to be:
(a) Quantity available on the market.
(b) Distance to market.
(c) Traditional market specifications for construction timber.

Best prices seem to be fetched by spruces and pines. These fetch 10–15 per cent more, generally, than species such as larch and Douglas fir, which are less abundant. However, in South Wales for example, where sawlogs of larch are more plentiful than those of other species, good markets have developed, and there is almost no price differential against larch. The volume of larch available in South Wales for harvesting in 1970 in sawlog sizes was estimated to be 25 000 cubic metres, in logs of 18 cm minimum top diameter and 3 m minimum length.

The effect of distance from markets is seen in the broad gradient of prices which are lowest in the north and west of the country, increasing towards the south and east. This price differential does not seem to depend on species. The gradient is nevertheless important.

Long-standing traditional market specifications can place limits on the use of home-grown timber. For example, the specification for timber used in house construction contained in British Standards Institute Code of Practice CP 112, permits the use of home-grown Scots pine, Douglas fir and the larches, but not Sitka nor Norway spruce nor other conifer species. However, the introduction of Mechanical Stress Grading (Anon, 1970) may lead to general acceptance of alternative standards, based on directly measured strength properties of graded timber of a much wider range of species.

6.52 Imported Timber

While in theory the prices of imported sawn timber of different species should be a useful guide to the different prices that home-grown timber might fetch, in practice it is difficult to determine the prices of similar quality timber of different species, since price is so much affected by grade, cross-section dimensions and lengths, and by availability. Quality and size is markedly more important than species, high quality joinery timber commanding a much greater price than constructional timber.

For the latter, the apparent distinctions are between Douglas fir, which commands a price a few per cent better than Sitka spruce and Western hemlock, both of which in turn are a few per cent more valuable than "Balsam fir" – a timber-trade name which covers all the north-west American *Abies* species (Western red cedar is not imported in construction timber grades). Thus, for imported timber there appears to be a broad though not precise agreement between prices and mean Nominal Specific Gravity.

6.53 Basis for Evaluating Prices of Minor Species

In assessing the likely prices to be expected for the four minor species in this study, the two principal effects to be taken into account are:
(a) basic timber properties as reflected in timber density, and
(b) the quantity of timber likely to be forthcoming in the future.

Both effects are evaluated in the following paragraphs.

6.54 Timber Density in Relation to Price

The nominal specific gravities of 10 conifer species are listed in Table 19. It is proposed to assume that future prices in due course will come to reflect those timber properties which are correlated with timber density. This assumption can be justified on the ground that dry fibre content which is the concern of the pulp miller, and strength which concerns any user of sawn timber, are both highly correlated with density. Mechanical stress grading of sawn timber, already introduced in Australia and under development in this country could hasten a price structure for sawn timber which more closely reflects strength than is currently the case.

Any variation in value due to features such as drought crack are additional to the differences given in Table 19.

6.6 Minimum Quantities for Effective Marketing

The present markets for wood processing and sawn timber differ in some important respect as regards species.

For many of the wood processing industries, species can be mixed, given sustained supplies in appropriate proportions. Prices for any one such species do not differ according to whether it is mixed or not. Sawn timber, on the other hand, is sold by name mostly species by species. If two species are mixed, the price for the mixed lots is generally lower. For example "hembal" – a commercial mixture of Western hemlock and various fir species, commands a slightly lower price than pure hemlock. (By exception, the North-western American true firs *Abies procera, A. grandis, A. lasiocarpa* and *A. amabilis* are marketed as 'Balsam fir'.)

Sections 6.1 to 6.4 have shown that there should be no difficulty in principle in disposing of sustained

TABLE 19
NOTIONAL FUTURE PRICE DIFFERENTIALS FOR TIMBER IN RELATION TO TIMBER DENSITY, RELATIVE ABUNDANCE AND MARKET PREFERENCE

Species	Nominal Specific Gravity (ex Table 14)	Weighting for Density	Weighting for Abundance/Market Preference for period 1975–1995	Notional Future Overall Weighting*
Scots pine	0·44	+10%	−5%	+5%
Corsican pine	0·43	+10%	−5%	+5%
Douglas fir	0·43	+10%	−5%	+5%
European larch	0·45	+10%	−5%	+5%
Japanese larch	0·43	+10%	−5%	+5%
Western hemlock	0·39	+5%	−10%	−5%
Sitka spruce	0·34	0	0	0
Norway spruce	0·33	0	0	0
Grand fir	0·31	−5%	−10%	−15%
Noble fir	0·32	−5%	−10%	−15%
Western red cedar	0·32	−5%	−15%	−20%

* A differential could well arise between pulpwood prices and sawlog prices. The higher dry fibre yield for pulpwood from some denser species could be entirely counterbalanced by the higher processing costs due to greater resin content and colour. These latter properties are no disadvantage for sawlogs, the price of which could well reflect density, features such as straightness, spiral grain, etc., being equal.

supplies of the smaller roundwood sizes of Grand fir or hemlock for pulpwood in Britain in mixture with spruces. Such pulping of mixed species is everyday practice in North-west USA and British Columbia. With Noble fir and Red cedar, staining and natural colour of the wood respectively may render the timber unsuitable for certain roundwood markets, but for the others, there is no evidence that there is any constraint that has to be overcome before either species is acceptable.

For sawn timber, the question is whether within a given radius related to transport costs there is a minimum scale of production which has to be achieved in order to create and sustain a local or regional market for home-grown sawn timber so as to obtain the best price for any particular species. It could be sufficient to ensure sustained supplies at the level of local industry, as it will depend principally on the skill of the individual sawmill manager in preparing for the market a product that is competitive in finish, seasoning, sustained supply, delivery, etc., with imported material or with the out-turn of other mills. However, even on this scale, a minimum sustained annual production approaching 20 000 cubic metres of sawlogs must be envisaged within an economic radius, (cp: the production regions described by Hummel and Grayson, 1968). Production at these levels would have to be backed by between 5 and 10 000 hectares of plantation per species, depending on the minimum size of the market and the mean yield class of the plantation.

Produce from plantations made on a lesser scale must perforce find its way to timber markets in mixed lots or in fluctuating quantities, for which the grower is likely to get poorer prices than for species for which there is an established market in home-grown timber.

6.7 Timber supplies from Existing Plantations

Table 20 shows the area in Forestry Commission Woodlands of the four minor species up to 1967. Comparisons with the acreage of the same species planted before 1950 emphasises that between 80 and 90 per cent of the total acreage has been planted in the last two decades. Table 21 gives the area of minor species in private woodlands up to 1965 and shows not only a similar high proportion of post-1950 plantations, but that the total acreage of all ages of minor species is less than 4 400 hectares.

These tables show that in most regions the existing plantations of the minor species make an insignificant contribution to timber supplies, and that during the next twenty years the bulk of the material coming forward will be the smaller sizes of roundwood.

Western hemlock is the only species which, with the existing area of plantation, could achieve a recognised, sustained market as a named home-grown sawn timber; however, this is only possible in North Wales (approx. 1 800 ha) and South-east England (approx. 1 900 ha) taking Forestry Commission and Private Woodlands together, and assuming that the planting rate of the last two decades continues.

Red cedar has been planted in South-west and South-east England (1 050 and 950 ha respectively) on a scale which, on the basis set out in the previous section, is sufficient only to start relatively small

TABLE 20
TOTAL FORESTRY COMMISSION AREA OF WESTERN HEMLOCK, RED CEDAR, GRAND FIR AND NOBLE FIR
UP TO 1967 BY CONSERVANCY; ALSO TOTAL BY COUNTRY UP TO 1950

Species	ENGLAND (Conservancy)					Total to 1967	Total to 1950
	NW	NE	E	SE (including New Forest)	SW (including Forest of Dean)		
				Hectares			
Western hemlock	681	399	291	1311	718	3404	323
Red cedar	328	80	375	865	250	1898	61
Grand fir	153	129	306	176	182	946	75
Noble fir	120	29	5	9	27	190	12

	SCOTLAND				Total to 1967	Total to 1950	WALES		Total to 1967	Total to 1950
	N	E	S	W			N	W		
			Hectares							
Western hemlock	167	46	188	647	1047	220	1657	1118	2774	140
Red cedar	20	6	32	133	191	43	239	254	492	44
Grand fir	81	65	131	102	378	124	519	236	755	103
Noble fir	94	20	399	344	857	88	433	152	585	15

local markets, and would have to be backed by substantial continuing planting programmes to build up supplies.

Neither Grand fir nor Noble fir are at present planted on a scale sufficient in any region to sustain a market for sawn timber.

From the foregoing, it is clear that none of the minor species will achieve the status of a "major species" in respect of timber which will be available on the market within the next 25 years. In Table 19 it has therefore been assumed that in addition to considerations based on density, the price to be expected for roundwood of the minor species will be 10–15 per cent lower than that of 'major' (abundant) species because of the relatively small volumes forthcoming.

TABLE 21
AREA OF WESTERN HEMLOCK, RED CEDAR, GRAND FIR AND NOBLE FIR IN PRIVATE WOODLANDS: BY COUNTRY UP TO 1950 AND TOTAL TO 1965
(Data from 1965 Sample Census)

	England		Scotland		Wales	
	Total to 1965	Total to 1950	Total to 1965	Total to 1950	Total to 1965	Total to 1950
	Hectares		Hectares		Hectares	
Western hemlock	1468*	50	102	38	325	Nil
Red cedar	1383**	125	32	29	93	Nil
Grand fir	206	84	147	30	72	8
Noble fir	81	40	93	68	2	2

*Approximately 600 and 560 ha of this in SW(E) and SE(E) respectively.
**Approximately 800 and 200 ha in SW(E) and SE(E) respectively.

Chapter 7
OVERALL COMPARISONS OF SPECIES

In the preceding sections, comparisons have been made of those individual aspects of silvicultural techniques, the incidence of pests and timber properties which affect the relative costs of, or benefits from, growing one or other major or minor species. In this chapter all such comparisons are brought together, expressed in financial terms discounted, where appropriate, at 5%. The differences which emerge are then related to observed relative rates of growth.

The bases for each set of comparative figures are given as far as possible; alternative figures can therefore readily be substituted, should other assumptions or generalisations be thought more appropriate than those made in this paper. Figures for those costs or revenues which in earlier tables have most appropriately been indicated as a range have here been simplified, using the median value of the range. However, the limitations of this practice are self-evident.

7.1 Costs
Table 5 (Section 2.1) summarises those costs of establishment which are affected by choice of species. The table shows that differences in these species-dependent costs are not great, though they favour the major rather than the minor species. All costs are discounted back to the year of planting.

All other costs are common to whatever species is planted and therefore require no mention here.

7.2 Revenues
The revenues from any given stand depend on its rate of growth and the prices obtained for different sizes of produce. In para. 7.21 and Table 22 discounted revenues are initially calculated, assuming the same price/size relationship for all species. In paras. 7.22–7.24 and Table 23, the differences in timber properties between species are then discussed and set out as well as possible differential losses due to conifer heart rot and to drought crack. These various influences on revenue and relative yields are summarised in Table 23 and Section 7.25.

7.21 Relationship Between Growth Rate, Yield Class and Revenue
The Forestry Commission Management Tables (Hamilton and Christie, 1971) show the expected yield of a full range of species and yield classes. If roundwood from intermediate and final fellings is valued strictly according to size and normal discounted procedure is followed, discounted revenues (DR) can be calculated for all species in terms both of cash and also as a percentage of the DR for Sitka spruce. Table 22 gives such revenues for species in this study.

TABLE 22
DISCOUNTED REVENUE* (5% DISCOUNT RATE) FOR VARIOUS SPECIES FOR YIELD CLASS 14 AND 18, AND THE PERCENTAGE DIFFERENCE IN DISCOUNTED REVENUE BETWEEN SITKA SPRUCE AND FOUR 'MINOR' AND FOUR 'MAJOR' SPECIES

Species	Yield Class	D.R. £	% of D.R. of SS	Yield Class	D.R. £	% of D.R. of SS
Sitka spruce	14	168	100	18	274	100
Western hemlock		123	80		212	75
Red cedar		148	83		238	85
Grand fir		180	—		270	95
Noble fir		139	83		222	80
Norway spruce		142	92		238	85
Douglas fir		201	125		305	115
Corsican pine		183	108		288	105
Japanese larch		267	167		—	
Scots pine		182	108		—	

Note:
*Discounted revenue has been calculated on the following basis:—
(i) a 15 per cent reduction on *all* management table figures to allow for gaps in crops, roads, etc.;
(ii) prices based on size as follows (stumpage price, p per cubic metre:—

	Diameter			
	15	25	35	45
	£ p			
	1·46	3·15	4·09	4·16

(Intermediate prices/sizes were also used but are not given; they can be estimated by interpolation if necessary.)

The percentage differences between the discounted revenue for any given species and that for Sitka spruce are only a little altered from those given in Table 22 if other yield classes or discount rates are chosen. For higher yield classes, the negative differences are reduced and positive differences increased by one or two per cent, and conversely for lower yield classes.

If higher discount rates are chosen, the extreme differences are exaggerated by one or two per cent, but the species relativities remain the same.

The differences in discounted revenue reflect the fact that minor species plantations commence thinning later and thinnings are smaller at any given stage in growth (Section 3.2).

7.22 Comparison of Discounted Revenues
Table 23 summarises discounted revenues for four

TABLE 23
COMPARISONS OF DISCOUNTED REVENUES AS AFFECTED BY SPECIES
All Figures Refer to YC 14
£ per hectare

Source of Revenue or of Revenue Adjustment	Western hemlock	Red cedar	Grand fir	Noble fir	Sitka spruce	Norway spruce	Douglas fir	Corsican pine	Scots pine	Jap/Hybrid larch
Discounted Revenue for Yield Class 14 assuming the same price/size for all species (ex Col 3, Table 22)	£123	£148	£180	£139	£168	£142	£201	£183	£182	£267
Change in Revenue if prices reflect timber density and abundance weighting shown in Table 19	−£6	−£30	−£27	−£21	0	0	+£10	+£9	+£9	+£13
Change (Loss) in Revenue due to Conifer heart rot (Fomes) (ex para 7.23)	0/−£25	0/−£15	0/−£5	0/−£5	0/−£25	0/−£25	0/−£12	0	0	0/−£25
Change (Loss) in Revenue due to multiple forking	0/−£5	0/−£10	0	0	0	0	0	0	0	0
Change (Loss) in Revenue due to Drought Crack (ex para 3.31)	0/−£5	0	0/−£12	0/−£25	0/−£5	0	0	0	0	0
Sum of Values in each column — Range	£82/117	£93/118	£143/153	£88/118	£138/168	£117/142	£199/211	£192	£191	£255/280
— Median	£100	£100	£148	£103	£153	£130	£205	£192	£191	£268

minor and six major species for YC 14, figures for Scots pine and Hybrid larch being added to those for the four major species for which full silvicultural details are given.

Timber Properties and Revenue
In Section 6.54 the case is argued that in the long run, timber prices for a given size and quality of log will principally reflect the density of the timber grown. The differences in value per unit volume in Table 19 have been related to the Discounted Revenues in Table 22 to give the figures shown in the second line of Table 23.

Loss in Revenue Due to Conifer Heart Rot (Fomes)
The loss in revenue due to *Fomes* has been obtained by taking the loss, expressed as a discounted percentage of production, from Section 5.1 and restating it in terms of Discounted Revenue for Yield Class 14, after adjustment in revenue for timber density. Thus, for Sitka spruce, a relatively susceptible species, a loss of up to 15 per cent of DR due to rot can be expected on fairly seriously infected sites. This corresponds to a loss in DR of up to £25. Noble fir being a relatively resistant species, is rated as liable to up to 5 per cent loss due to *Fomes*, i.e. a loss of up to £5 DR, which is up to £8 less than for Sitka spruce. Values for other species have been derived similarly.

Losses in Revenue due to Drought Crack
These figures are derived from Section 3.31 where the conclusion is reached that drought crack could be of economic significance in crops of Noble fir, Grand fir and Sitka spruce.

7.3 Combined Costs and Revenues
Discounted costs and revenues are brought together in Table 24. The line 3 of the figures in the table indicates differences in the value of species *at the*

TABLE 24
COMBINED SPECIES-DEPENDENT COSTS AND REVENUES
(All Crops YC 14)

£ per hectare

	WH	RC	GF	NF	SS	NS	DF	CP	SP	JL
					SPECIES					
					EXISTING CROPS					
Median Revenues from Table 23	£100	£100	£148	£103	£153	£130	£205	£192	£191	£268
Median Costs from Table 5	£80	£100	£120	£137	£70	£94	£75	£82	£57	£66
Revenue less Costs (existing plantations)	£20	£0	£28	−£34	£83	£36	£130	£110	£134	£202
					FUTURE CROPS					
Median Increase in D.R. due to improved seed (future plantations)	£17	£7	£0	£0	£45	£22	£22	£17	£45	£45
Revenue less Costs − future (improved plantations)	£37	£7	£28	−£34	£128	£58	£+52	£127	£179	£247

same rate of volume production in current plantations. The differences are applicable to crops currently growing; they indicate that all four minor species currently carry a substantial financial handicap which can only be offset by higher growth rates compared with major species. These figures underestimate the differences in plantations which could be planted in the next two decades, at least on more fertile sites, since no account has been taken of the increases in volume production in plantations made with plants raised from genetically improved seed. Potential increases in yield from this source (Table 13, Section 4.3) have been quantified, valuing an improvement of one yield class at £45 DR and are given in line 4 of Table 24; line 5 corresponds to line 3 but taking into account these potential improvements. The handicap to the minor species mentioned earlier in this paragraph is substantially increased in the context of planting during the next two decades when unimproved sources of minor species will be available for sites in competition with at least some 'improved' seed sources of the major species.

Tables 22–24 illustrate that the biggest influence on profitability are:
(1) early growth rate, species which are slow to come into production being less valuable;
(2) timber density, assuming that in the long term, prices will reflect density;
(3) abundance of wood supplies and hence effect on market prices;
(4) potential increases in future plantations due to improved seed.

All these factors operate to the disadvantage of the minor species. The first two reflect species characteristics but the fourth factor, and to a limited extent the first two also, could be expected in the long term to be much less of a disadvantage if intensive breeding programmes were set in motion. The third factor will only disappear if, either much larger areas of plantations of minor species are created, or a major change occurs in the marketing practice for sawn timber.

The tables, being based on broad average values obscure particular differences that may apply in particular circumstances. Thus, the *relative* value of Grand fir on sites heavily infected with *Fomes* butt rot could be £25 DR better than is shown in Table 24. Similarly, when established under shelter, the discounted revenue from Western hemlock and Grand fir could be enhanced by up to £25 per hectare over the broad average values in Table 24 on account of lower establishment costs and rapid growth associated with shelter.

If the difference in yield of one yield class is valued at £45 DR, Table 24 provides the basis for estimates of what rates of growth have to be achieved by different species for any pair of species to be of equal value. Some estimates are given in Table 25. These show the number of yield classes *more* which the minor species named has to yield in order to equal in value the major species with which it is being compared on any given site.

7.4 Combination of Growth Rates and Valuations
On the one hand, the four "minor species" with which

TABLE 25
NUMBER OF ADDITIONAL YIELD CLASSES PRODUCTION NECESSARY IF MINOR SPECIES ARE TO EQUAL MAJOR SPECIES IN VALUE**

	SS	NS	DF	CP	SP	HL/JL
Western hemlock						
existing	1½	½	2½	2	2½	4
future*	2	½	2½	2	3	4½
Red cedar						
existing	2	1	3	2½	3	4½
future*	3	1½	3½	3	4	5½
Grand fir						
existing	1	½	2½	2	2½	4
future*	2	½	3	2	3½	5
Noble fir						
existing	2½	1½	3½	3	3½	5
future*	3½	2	4	3½	4½	6

Notes:
*Assumes 'improved' seed (Section 2.39).
**All these figures are subject to a margin of uncertainty due to cost differences within the known range of practice and response to site conditions (Tables 5 and 23).
The margin is at least ± ½ yield class for all species and up to ± 1 yield class for Noble fir and Norway spruce.

this bulletin deals are shown in Section 3.1, Table 9 and Figures 1–18 to produce greater volumes of timber on the better forest sites. On the other hand, the discounted value of such wood is shown in Section 7.3 to be less, yield class for yield class, than for the major species.

The extent to which the species' greater volume production offsets their lower discounted value is outlined below and summarised in Table 26.

7.41 Western hemlock

In relation to Sitka spruce, the lower discounted revenues listed in Table 23 nullify almost all the advantages of greater rates of growth illustrated in Figure 9. There is no clear overall case for preferring Western hemlock to Sitka spruce; however, if hemlock is to be preferred, it should be on infertile (dry) but sheltered sites.

By contrast, hemlock appears more profitable than Norway spruce on the large majority of sites.

In comparison with Douglas fir, the distinction previously noted between Scottish WH/DF comparisons and those elsewhere in Britain leads to the conclusions that, in so far as the Scottish data is representative, there is no case for preferring one species over the other in Scotland. For England and Wales, the data is scattered but there is an indication that hemlock and Douglas fir are about of equal value on YC 16 Douglas fir sites, hemlock having the advantage on the better sites.

After allowances are made for the lower value of hemlock, the value of Corsican pine is equal to or greater than that of hemlock on the limited number of sites where both species were compared.

7.42 Red cedar

Red cedar is more responsive to site condition in

TABLE 26
SUMMARY OF COMPARISONS OF VALUE OF CROPS BY SPECIES AND SITE

	Sitka spruce	Norway spruce	Douglas fir	Corsican pine
Western hemlock	WH ≤ SS on most sites (WH > SS on sheltered infertile sites)**	WH > NS on most sites down to NS sites YC 12/14	(For Scottish sites, WH ≤ DF)** for others WH > DF on DF sites YC 16 and better	(WH < CP)**
Red cedar	RC < SS above 150 m and in Scotland. RC ≥ SS below 150 m in England and Wales	(RC = NS)**	RC ≥ DF on DF sites YC 18 and better	(RC ≤ CP on free draining soils; RC ≥ CP on heavy soils with impeded drainage)*
Grand fir	GF ≥ SS on SS sites YC 18 and better	GF > NS on almost all sites where compared (NS sites YC 14 and better)	DF ≥ GF	GF ≥ CP on CP sites YC 14 and better
Noble fir	NF < SS	NF ≤ NS	(NF < DF)**	(NF < CP)**
	Changes due to relative valuation have a decisive influence to the disadvantage of Noble fir.			

*B < A
 A > B } indicates that A is greater than B in value.

B ≤ A
 A ≥ B } indicates that A equals or is greater than B in value.

**Text in brackets is supported by very limited data and must be treated with more than usual caution.

Britain than most other species and reflects this in its relative growth. It is at a considerable disadvantage in terms of value and appears to be able to offset this disadvantage only in the southern half of Britain on the best lowland sites in competition with Sitka spruce and Douglas fir, and on heavy impeded soils in comparison with Corsican pine.

7.43 Grand fir

Relating results in Section 3.1 and 7.3, Grand fir is of value equal to or greater than Sitka spruce on sites where Sitka spruce is growing at Yield Class 18 or better. The 1967 Census data showed approximately 7 000 ha of Sitka spruce in Forestry Commission plantations in Yield Class 18 or better out of a total of 220 000 ha under Sitka spruce (i.e. 3 per cent of total area is YC 18 or better).

In relation to Norway spruce, even after all adjustments have been made, Grand fir appears to be appreciably more profitable on all sites studied. These are, however, the more productive Norway spruce sites – three-quarters of the sites selected for comparisons with Grand fir were sites Yield Class 14 or better which represent roughly one-fifth of the Forestry Commission plantations of Norway spruce, i.e. an area of 16 000 hectares.

The advantage of Grand fir over Douglas fir after adjustment, is less clear-cut than over either of the spruces. Over the range of yield classes studied, there is a wide scatter of points without any clear indication of the important underlying factors. On 10 out of 19 sites, Douglas fir has the advantage; on 7, Grand fir is better and on 2 there is no difference. However, the sites where one or other species is better cannot at present be defined adequately in terms of site factors or yield class.

Corsican pine production equals in value that of Grand fir for pine Yield Class 14. On better sites Grand fir has the advantage. (28 per cent of Corsican pine plantations are Yield Class 14 or better, i.e. 9 000 out of 32 000 hectares.)

7.44 Noble fir

The value of Noble fir is the least of any major or minor species; it loses on almost every count. It is relatively expensive to establish, thinnings are delayed and revenues are therefore discounted over longer periods, timber is of low density and the species is unimproved. The species little more than holds its own in growth rate compared with other species and when its lower value is taken into account, there are no sites where Noble fir is more profitable than one or other of the major species.

7.45 Scots pine

Scots pine was included in only 7 comparisons (illustrated in Figure 17). As would be expected, it was inferior to the minor species in production on all sites but differences in value cancel out the minor species growth advantages on many sites. The greater production of the southern Red cedar plot relative to Scots pine, compared with the two northern ones, is consistent with the pattern for Red cedar already noted. It is perhaps more unexpected that hemlock has done no better than Scots pine on the three sites in Lincolnshire and Yorkshire. The one comparison between Noble fir and Scots pine was outstandingly in favour of the former.

7.46 Larches

There were altogether 19 comparisons between Japanese, Hybrid or European larch and one or other of the minor species, the results for Japanese larch being illustrated in Figure 18.

On all sites, the minor species produced substantially greater volumes. However, the larches are the most valuable *yield class for yield class* of any commercial species. Western hemlock has to grow at a rate of about 4 yield classes higher than larch to equal it in value. Grand fir and Red cedar have to yield 5 yield classes more while for Noble fir the difference is even greater. Fig. 18 shows that the minor species on the better sites can overcome even this handicap. However, on the less productive sites, improved Hybrid larch could remain as the most profitable species.

Chapter 8
FUTURE USE OF SPECIES

Because of the long time taken to grow timber crops in Britain, any decision to change the ratio of species planted can have no effect on timber production for at least 15 years and only in the following two decades will the cumulative effect of present decisions be felt. (The effect on past decisions on planting in terms of areas of existing plantations are given in Tables 20 and 21, Section 6.61.)

Quite clearly from the preceding sections, the minor species can only be considered as potentially viable for the more productive forest sites in Britain. On the basis of the Forestry Commission's present land acquisition pattern, the areas of previously unplanted land which would be best planted with Grand fir or Western hemlock and which are likely to be forthcoming in the future, are extremely limited. On the evidence of recent years, the total area of new Forestry Commission acquisitions more suitable for these than other species is likely to be between 200 and 400 hectares per annum dispersed in small blocks over a large area.

The area of forest due annually for regeneration will, however, increase rapidly over the next two or three decades and is likely to include substantial areas well suited to the minor species.

8.1 Potential Use of Western hemlock

Western hemlock is intermediate in position in relation to major species. On the available evidence, hemlock should not be planted if Sitka spruce can be grown on a site. Thus, while hemlock could profitably replace Norway spruce on very many sites, it would only be logical to do so on sites unsuited for Sitka spruce. Similarly, when considering replacements for larch, hemlock should be considered only after both Sitka spruce and Grand fir have been considered and rejected. Because it is more tolerant of infertile soils, hemlock could find a place on mineral soil sites too dry for Sitka spruce and too infertile for Grand fir. In the southern half of Britain, it is likely on such sites to take second place to Corsican pine. However, in East Scotland and North-east England, it could be a preferable alternative to Scots or Lodgepole pine, especially where additional shelter from an overcrop can be provided.

Because of the areas already existing in Wales and the South of England, quantities of home-grown hemlock sawlogs can be expected in about 10 years' time. In these areas, the decision has to be made whether to plant Grand fir on a scale that will sustain a sawlog market starting in about 2000 A.D., which if the present assumptions still apply would be the most profitable course, or to plant hemlock in the hope of sustaining, again in 2000 A.D. and later, a market already established.

Because hemlock is second choice over so many sites, no attempt has been made to quantify the area of land potentially plantable to hemlock.

8.2 Potential Use of Western Red Cedar

The species appears to be at its most competitive on the heavier texture lowland soils in southern England. Elsewhere, it is not outstanding and in Scotland it often fails. About 2,000 hectares of young plantations exist in private and Forestry Commission woodlands, divided about equally between South-west and South-east England. While there is likely to be a ready market for high-quality sawlogs, because of the colour of the wood and its chemical constituents, small-dimension Western red cedar is not acceptable for most of the pulpwood processes currently existing in Britain, nor for the manufacture of most other processed wood products. This reason is probably sufficient to lead to a decision to plant other species (Grand fir or hemlock) on sites suitable for Red cedar. This is essentially a southern problem; there is no evidence to justify any large-scale usage in Scotland or northern England.

TABLE 27
RELATIONSHIPS BETWEEN LOCAL YIELD CLASS OF GRAND FIR AND MAJOR SPECIES

Species	Regression constant: (a)	coefficient (b)	Correlation coefficient	Significance level	No. of pairs of plots
Sitka spruce	0·80	1·19	0·64	0·01	16
Norway spruce	8·8	0·75	0·76	0·001	19
Douglas fir	7·8	0·83	0·40	Not Significant	19
Corsican pine	3·0	1·23	0·73	0·001	11

Regression equation data are used as follows:
    Grand fir LYC = Appropriate constant (ex col a) + (observed LYC of species to be replaced x appropriate regression coefficient ex col b).

TABLE 28
ESTIMATED AREAS OF FORESTRY COMMISSION WOODLANDS REACHING THE AGE OF 50 (140 FOR BROADLEAVES) IN THE NEXT 20 YEARS AND WHICH ARE POTENTIALLY REPLACEABLE BY GRAND FIR

hectares

		CONSERVANCY: ENGLAND				
		NW	NE	E	SE (including New Forest)	SW (including Forest of Dean)
Sitka spruce	YC 18+	80	180	—	36	320
Norway spruce	YC 14+	340	520	32	60	700
Douglas fir	YC 16+	360	40	32	220	450
Corsican pine	YC 14+	40	—	1500***	425*	120*
Larches	YC 12+	100	20	—	30	260
Pre 1900 Broadleaves	YC 6+**	200	40	160	900	800

		CONSERVANCY: SCOTLAND				CONSERVANCY: WALES	
		N	E	S	W	N	S
Sitka spruce	YC 18+	590	220	360	620	120	320
Norway spruce	YC 14+	650	300	700	930	770	460
Douglas fir	YC 16+	590	80	80	60	300	160
Corsican pine	YC 16+	—	—	—	—	—	—
Larches	YC 12+	140	120	80	160	160	440
Pre 1900 Broadleaves	YC 6+**	—	60	20	80	120	320

Notes: The figures here are extracted from Conservancy records based on conclusions set out in Table 26. For example, Table 26 indicates that GF is as profitable or more profitable than SS on SS sites growing at YC 18 or better. The first line of Table 27 tabulates the area of plantations planted before 1940 and shown in Conservancy records as being in YC 18 or better.
  *Includes 160 ha of SP YC 14 in SE(E) and 60 ha in SW(E)
  **Should be assumed not to be available for conversion to any conifer
  ***Shelter essential in most of this area.

8.3 Potential Use of Grand fir

Grand fir is the one minor species which, because of its very high productive rate on the best sites, can overcome the handicaps set out in Section 7.4 and, on the assumptions made, can be grown more profitably on such sites than any other species. In present circumstances sites for Grand fir can be expected to be found chiefly among older stands rather than on sites currently or recently afforested.

Any estimate of the area of forest which might in due course be replaced with Grand fir depends on the reliability of the correlation between local yield class of the major species to be replaced and Grand fir. Correlations and regression equations derived from the data illustrated in Figures 1, 5, 9 and 13 are given in Table 27.

These show that forecasts of Grand fir production on sites currently under spruce or Corsican pine will be more reliable than similar forecasts for sites under Douglas fir and Japanese larch. For Japanese larch, local yield class is not a sensitive basis for the prediction of Grand fir productivity; however, this does not invalidate the more general conclusion that Grand fir out-yields Japanese larch substantially on the more favourable sites. Given this limitation, Table 28 lists those areas of Forest Commission woodlands which are likely to fall due for replacement (assuming replacement at the age of 50) and are growing at rates which indicate they might be most profitably replaced by Grand fir. If a "normal" production area of Grand fir (in the technical sense) is to be achieved in any region and a level of sawlog production anywhere near 20 000 cu m/annum, maintained, between 50 and 100 ha must be planted annually. If the areas listed in Table 28 were all to be planted with Grand fir, many Conservancies could plant the species at the rate of 50 ha per annum, but none could plant at 100 ha annually without making inroads into substantial areas of mature hardwoods or extending planting on to sites where Grand fir would not clearly be at an advantage.

Because of the limited area of existing plantation of Grand fir, there are greater elements of uncertainty and speculation in using the species extensively than for most other species. On the credit side are its undoubted ability to grow fast, its resistance to *Fomes* butt rot and its shade tolerance. On the debit side are its low timber density, and absence of an existing sawn timber market in Britain, risk of damage and associated 'rotholz' formation linked with the complex of drought, exposure and *Adelges piceae*, and its susceptibility to late spring frost.

It can be argued that it is in the western half of Britain that the prospects of high growth rates are most certain and the risk from drought are least, and that it is here rather than in the east that the species should be grown. It is quite clear, however, that until positive decisions are taken to increase substantially the planting programme in one or more regions, Grand fir will remain a species of no practical importance in British forestry.

Two further considerations could affect the large-scale use of Grand fir. Firstly, if such a use is at the expense of a second species, the area of which itself is limited, this area could also be reduced to below what is necessary to sustain a market. Thus the area of Douglas fir plantation in several regions is on the low side, assuming an area of forest of 2 500 to 4 000 ha is required to maintain a regional market. In such regions, the choice might lie between continuing with a species such as Douglas fir and converting Douglas and all other suitable stands to Grand fir.

The second consideration is that in eastern England and wherever else there is a severe danger of late spring frosts, Grand fir can currently *only* be established under shelterwood. (Section 2.2, page 6.)

8.4 Potential Use of Noble Fir

There appears to be no case for any large-scale planting of this species, because of its slow early growth and low density timber.

8.5 Amenity and Recreation

Whatever may be decided about the large-scale regional use of given species, each of the minor species has its characteristic appearance which can be utilised to provide visual contrast and diversity in strategic places. Indeed, Western hemlock and Noble fir are outstanding in this respect. The graceful pendulous foliage of hemlock contrasts well with the stiffer habit of most other conifers, while Noble fir by colour and its conspicuously whorled branch pattern forms a most attractive contrast at the other extreme.

Grand fir could have a particular place in groups in sheltered localities where the public may wish to see big trees.

These decorative properties of minor species will probably ensure them a small but nevertheless increasingly important role in many forests, independent of policy for timber production.

SUMMARY

(With cross references to relevant Bulletin sections)

A study has been made during the last four years of the present status and future potential of four 'minor' species – Western hemlock, Western red cedar, Grand fir and Noble fir.

Crops of these species planted before 1950 and now at the pole crop or later stage of growth, are restricted in extent. Successful crops are largely to be found on the better soils in lowland or sheltered areas. (Section 1.3, page 3). Table 29A summarises some of the more important features of the different species.

Establishment and Early Growth

Costs of establishment differ according to species; the minor species are the more expensive to establish but the differences are not excessive. The survival and early growth of hemlock, Red cedar and Grand fir is markedly better under shelter than on the open hill. (Section 2.1, page 5.)

Grand fir and, to a lesser extent, Western hemlock are susceptible to late spring frost. This is a limitation of importance in areas such as East Anglia, where the risk of damage by frost is high, protection by an overstorey may be essential if these species are to be successfully established. (Section 2.2, page 6.)

Growth rates after planting differ widely, Western hemlock on many sites making rapid early growth whereas Noble fir and Grand fir are slower than any other commonly planted species (except where species are in check). Noble fir is almost always slower than Grand fir. The best early growths of hemlock has usually occurred under some form of shelter from an overcrop. (Section 2.3, page 6.)

Growth in Pole Crops and Later

Once established, growth of Grand fir, Western hemlock and Western red cedar is more rapid than that of most major species on the more fertile soils. On less productive sites, the pattern is not consistent. The differences in growth rate on the better sites are the equivalent of between 1 and 3 yield classes mostly, but more in extreme cases. (Section 3.1, page 9.)

For any given yield class, crops of all major species are ready for thinning three to five years earlier than other species. Thinning of Grand fir and Noble fir is delayed because of slow early growth, while thinning of hemlock is deferred because of slender form and consequent relatively low volume per pole cut. In later stages, hemlock especially and to a lesser extent other minor species carry more stems per acre at a given height than other species. Discounted revenue from minor species stands is depressed on both counts. (Section 3.2, page 46.)

Of various defects noted, drought crack was found to be most severe in Noble fir and is likely to have a definite adverse effect on revenues from stands severely affected. Grand fir was the species next most frequently found with cracks but the cracking was less severe and is likely to have only a minor effect on utilisation. The incidence of drought crack in other species was low. Other stem defects, i.e. fluting, buttressing, sweep, wavy stem all occurred to different degrees but none were severe enough to have any clear quantifiable effects on revenue. (Section 3.3, page 47.)

No evidence was forthcoming that any one species was more susceptible to windblow than any other. (Section 3.4, page 51.)

Use of Improved Seed in Plantations to be Made in the Next Two Decades

Potential increases in yield due to seed of improved genetic quality are greatest for larch and Scots pine and least for the minor species. (Section 4.1, page 52.)

Pests

The minor species show important differences in resistance to only one fungal pest, *Fomes* butt rot. Differences in resistance to Honey fungus have been observed but are not important. Grand fir stemwood is substantially more resistant to *Fomes* rot and very little loss of timber has been found. However, for Western hemlock, losses on heavily infected sites can be relatively severe, corresponding to a drop in the yield equivalent to up to a yield class. Noble fir is close to Grand fir in resistance to rot while most major species other than pines are intermediate between Grand fir and hemlock. While the area of ground at high risk from *Fomes* is small relative to the total area of Forestry Commission plantation, part of this land is on good soils where both Grand fir and Western hemlock could be expected to do well. (Section 5.1, page 59).

Only one insect presents any threat – *Adelges piceae* which can cause the formation of 'rotholz' in Grand fir. At present, this pest is of minor importance. (Section 5.4, page 60.)

There are no animal pests which constitute any more of a risk to minor species than to any other species. (Section 5.5, page 62.)

Timber Properties

Wood properties of home-grown Western hemlock give no reason to question the future disposal of good quality sawlogs or pulpwood of this species. The density of samples tested has been only slightly less than the average density values of hemlock imported from Canada or the United States. It is

TABLE 29A
SUMMARY OF PROPERTIES OF SPECIES

Species	Early Growth Rate	Overall Growth Rate	Shade Tolerance	Resistance to late Spring Frost	Performance on Dry Infertile Soils (lowland)	Performance on Exposed Sites	Performance on Peaty soils	Resistance to *Fomes* Rot	Resistance to Drought Crack	Relative Timber Density
Western hemlock	very good	good – very good	very good	moderate	good	moderate	moderate/poor	poor	good	intermediate
Western red cedar	moderate	good	good	good	moderate	poor	very poor	moderate	very good	low
Grand fir	poor	very good	very good	poor	poor	poor	very poor	good	poor	low
Noble fir	very poor	good	very good	moderate	poor	good	moderate	good	very poor	low
Sitka spruce	good	good	moderate/poor	moderate	poor	very good if rainfall adequate	very good	moderate	moderate	intermediate/low
Norway spruce	moderate	moderate/good	moderate	moderate	poor	moderate	moderate	moderate	moderate	intermediate/high
Douglas fir	good	good – very good	moderate	moderate	moderate	moderate	poor	moderate	good	high
Corsican pine	moderate	moderate/good	poor	good	v. good if warm	moderate/good if warm	poor	good	very good	high
Scots pine	moderate	poor	poor	very good	good	moderate/good if dry	moderate	good	very good	high
Japanese larch	very good	poor	poor	moderate	good	good	moderate/good	moderate	very good	high

slightly higher in density than home-grown Sitka spruce but gave slightly lower pulp yields. Hemlock is well established as a pulpwood in mixture with Sitka spruce in NW America. (Section 6.1, page 64.)

Home-grown Western red cedar, while similar in properties to imported timber, could present problems if marketed on a substantially larger scale than at present. For small roundwood, peeling currently presents difficulties, since existing designs of peelers become choked by stringy bark. The peeled timber is at present only acceptable as pulpwood for board manufacture. Its colour and the presence of certain chemical constituents not commonly found in other species, make it unsuitable for ground wood pulp and for chipboard and similar particle boards. Red cedar can be made into high quality Kraft pulp but this requires specialised mills dealing predominantly with the species, any such mill requiring several hundred thousand cubic metres of timber annually to be viable. There is no prospect of such scale of production in Britain in the present century. Disposal of sawlogs presents no obvious problems. (Section 6.2, page 65.)

Grand fir timber is among the least dense of any of the species commonly raised in British forests. However, its sawing and strength properties suggest it could be used for similar purposes to those for which home-grown Sitka spruce is suitable. Grand fir can be pulped in mixture with Sitka spruce and other white woods, though with a small reduction in yield and in strength of pulp produced. (Section 6.3, page 67.)

Noble fir is similar in density and most other properties to Grand fir. However, while clear specimens are somewhat stronger, drought crack, excessive taper and stain of heart wood can all reduce the out-turn in conversion of sawlogs. Little is known of Noble fir as a pulpwood but it is inferred that it could be used in the same way as Grand fir. (Section 6.4, page 68.)

Timber supplies from existing plantations are likely to make little impression on future markets, except for Western hemlock in Southern England. To ensure sufficient timber is available to sustain regional market, a minimum sustained production of 20 000 cubic metres of sawlogs must be envisaged within an economic radius. (Sections 6.5–6.7, pages 70–72.)

Costs and Revenues
Species-dependent costs of establishment are slightly higher for the minor species than major species. (Section 7.1, page 73.)

Discounted revenues from different species for the same yield class differ substantially because of the effects of early growth on the timing of thinnings and hence revenues. Revenues from the four minor species are all lower than those of any of the major species. If the assumption is made that future timber prices will more closely reflect timber density than at present, the revenues for Western hemlock will be enhanced slightly relative to other species. Taking into account potential losses by *Fomes* heart rot works to the advantage of Grand and Noble fir. (Section 7.2, page 73.)

Combining costs and revenues, if all the assumptions in previous sections are accepted, the various minor species suffer an economic handicap equivalent to between 1 and $2\frac{1}{2}$ yield classes in comparison with most major species. (Section 7.3, page 74.)

When the handicaps are related to growth on similar sites, Grand fir is the only species clearly at an advantage, compared with all other species. However, this is only so on the more productive sites. Noble fir is nowhere more profitable. Western hemlock comes out with neither an outstanding advantage nor disadvantage. Western red cedar is at an advantage only on lowland heavy soils. (Section 7.4, page 75.)

Grand fir is at its greatest advantage in terms of volume production on fertile soils in areas of high rainfall. There is some doubt as to the species' ability to survive the combined effects of drought and insect attack on dry or exposed sites in low rainfall areas. The maximum areas which could profitably be converted to Grand fir over the next twenty years are estimated to range from 500 to 1600 hectares per conservancy. However, unless there is a major increase in the area of plantation under Grand fir, it will remain a species of no commercial importance in British forestry. (Section 8.3, page 79.)

Because of the greater uncertainties, no figures have been estimated for areas that might be converted to other species. However, Western hemlock is most competitive on dry brown earth and intergrade soils which are not exposed. On such sites, it is likely to be more profitable than Scots or Lodgepole pine. However, if Corsican pine will grow to maturity on such sites this species is likely to give higher returns than hemlock. Similarly, where Sitka spruce or Grand fir can be grown on a large scale, they also will usually give higher returns than hemlock. (Section 8.1, page 78.)

Western red cedar has succeeded well only in the southern half of Britain. Its more extensive use is risky because of uncertainties of utilization of small sizes of roundwood. It is at best advantage on the more fertile, heavy textured lowland soils. (Section 8.2, page 78.)

Noble fir is nowhere sufficiently profitable to justify its commercial use. (Section 8.4, page 80.)

At present, only Western hemlock and Red cedar have been planted on a sufficient scale to begin to form the basis for any regular substantial supply of sawn timber. Even this only applies to Wales and Southern England, and in both of these areas further planting is essential if markets are to be developed. If Grand fir is to be planted on a commercial scale in future, in many regions, a choice may have to be made between Grand fir, hemlock and Douglas fir as there is unlikely to be sufficient suitable land to support crops for markets for more than one, or at the outside two, of these species.

REFERENCES

ANDERSON, H. E. (1955) — Climate in South-east Alaska in Relation to Tree Growth. *Alaska Forest Centre Stn. Paper* 3.
ANON (1970) — *Rep. Forest Prod. Res.* 1970 10
ANON (1969) — Kiln drying schedules. *Building Research Establishment, Princes Risborough Laboratory Technical Note* 37.
ANON (1967) — *Fomes annosus*. A fungus causing butt rot, root rot and death of conifers. *Leafl. 5 For. Comm. Lond.*
ANON (1965) — Silvics of Forest trees in the United States. *Agricultural Handbook* 271. US Dep. Agric., Forest Service.
ANON (1965) — Seed Identity Lists. *Res. and Devel. Paper* 29 *For. Comm. Lond.*
ANON (1955) — Wood Handbook. *Agriculture Handbook* 72. US Dep. Agric. Forest Prod. Lab.
BRAZIER, J. D. (1973) — Wood properties of minor softwood species. *Forestry Home grown Timber.* 2(5): 27–28, (6): 37–38
BROUGHTON, J. A. H. (1962) — Properties of 30–37 year old Sitka spruce timber. *Forest Prod. Res. Bull.* 48
BURDEKIN, D. (1970) — Death and Decay Caused by *Fomes annosus*. *Rep. For. Res. Lond.*, 1970. 114–115.
BUSBY, R. J. N. (1962) — Species and forms of the Silver fir Adelgid in Scotland. *Scott. For.* 16: 243–254.
BUSBY, R. J. N. (1964) — A new Adelgid on *Abies grandis* causing compression wood. *Q. Jl. For.* 58: 160–162.
CARTER, C. I. (1971) — Conifer Woolly Aphids (Adelgidae) in Britain. *Bull. For. Comm., Lond.* 42.
CASEY, J. P. (1960) — *Pulp and Paper*, Vol. I, p. 69.
DAY, W. R. (1954) — Drought crack of conifers. *Forest Rec., Lond.*, 26
DOERKSEN, A. H. and MITCHELL, R. G. (1965) — Effects of the Balsam Woolly Aphid upon wood anatomy of some Western true firs. *Forest Sci.* 11: 181–188.
FOWELLS, H. A. (1965) — Silvics of Forest Trees of the United States. *Agricultural Handbook* 271. 1–762.
GARDNER, J. A. F. (1963) — The chemistry and utilisation of Western red cedar. *Canada Dept. of Forestry Publication*, 1023.
GAVELIN, G. (1966) — *Paper Trades Journal.* 150 (17): 80–82.
GREIG, B. J. W. (1969) — On drought crack in Grand fir. *Supplement to Timber Trades Journal, 29 March 1969.* pp. 26–27.
HAMILTON, G. J. and CHRISTIE, J. M. (1971) — Forest Management Tables (Metric). *Booklet For. Comm., London.* 34.
HOLSTENER-JØRGENSEN, H. (1973) — Fertilising experiments in Abies nobilis decoration-greenery stands. *Det Forstlige Forsoeksvaesen: Danmark* 33(3): 289–301.
HOWELL, R. and NEUSTEIN, S. A. (1966) — The influence of geographic shelter on exposure to wind in Northern Britain. *Rep. Forest Res., London*, 1964/5. 201–3.
HUMMEL, F. C. and GRAYSON, A. J. (1968) — Future wood supplies in Great Britain. *Forest Rec. Lond.* 68: 4–16.
GREENWOOD, J. H. F. (1973) — Wood working Tests on the Minor Species. Unpublished report. Princess Risborough Laboratory.
JOHNSTON, D. D. (1966) — The specific gravity and moisture content of conifers. *Scott. For.* 20 (4): 255–260.
KRAJINA, V. J. (1965) (Ed.) — *Ecology of Western North America.* Vol. I. Dept. Botany, Univ. Brit. Columbia, Canada.
KRAJINA, V. J. (1969) — Ecology of Forest Trees in British Columbia. *Ecology of Western North America.* 2(1) 1–146.
LAVERS, G. M. (1969) — The strength properties of timbers. *Forest Products Research Bulletin* 50 (Second edition).

LINES, R. (1970) — Note on provenance of coniferous forest tree seed for use in Scotland. *Scott. For.* 24 (1): 10–13.

LINES R. and MITCHELL, A. F. (1968) — Provenance of Sitka spruce. *Rep. Forest Res. London 1968*, 67–69.

MAYHEAD, G. M. (1973) — Some drag coefficients for British forest trees derived from wind tunnel studies. *In press.*

PEACE, T. R. (1938) — Butt rot of conifers in Great Britain. *Q. Jl. For.* 32, 81–104.

PEACE, T. R. (1962) — (Section on *Armillaria mellea* in) *Pathology of Trees and Shrubs with Special Reference to Britain.* Clarendon Press, Oxford.

PRIEST, D. T. (1973) — The Sawmilling properties of four minor Home-grown Softwoods. Unpublished report. Princes Risborough Laboratory.

PYATT, D. G., HARRISON, D. and FORD, A. S. (1969) — Guide to Site Types in Forests of North and mid-Wales. *Forest Rec., Lond.* 69.

WEATHERN, J. D. (1959) — Western red cedar utilisation. *Forest Prod. J.* 9 (9); 308–313.

WOOD, R. F. (1955) — Studies of North-west American Forests in Relation to Silviculture in Britain. *Bull. For. Comm. Lond,* 25.

WOOD, R. F. and NIMMO, M. (1962) — Chalk Downland Afforestation. *Bull. For. Comm., Lond.* 34.

APPENDIX I
CROP, SITE AND SOIL DATA

Data

This report is based on:
(i) Crop and site data collected for the purpose by means of special surveys, during 1967 and 1968, of plantations established in 1950 or earlier.
(ii) The results of laboratory studies and analyses of plant and soil samples collected during the surveys.
(iii) Summarised data from over 150 experiments in which species have been growing under strictly similar conditions.
(iv) Information provided by each Forestry Commission Conservancy on the then current practice for planting and tending.
(v) Data from the Management Services Division of the Forestry Commission on growing stock by Conservancy, age and yield class; estimates of likely future felling schedules, etc.
(vi) Data from Harvesting and Marketing Division and from the Princes Risborough Laboratory of the Building Research Establishment on timber properties, and trends in utilization of timber.
(vii) Meteorological data provided by the Meteorological Office.

Site Selections

When selecting stands for the special survey in 1968, the country was stratified on a crude climatic basis using, with minor adjustments, the 11 Meteorological Office regions used in the Meteorological Office Monthly Weather Reports. These regions correspond quite well with Conservancy boundaries except in the Midlands of England and South Scotland. Table 29B shows the number of plots by species in each of the Meteorological Office regions.

Where possible in each region, for each minor species, two stands of high, two of medium, and two of low yield were chosen, basing yields on the results of a preliminary survey. On this basis, 66 stands of each minor species should have been selected, but in some regions there were too few suitable stands. While some deficiencies could be made up from other regions, Noble fir, and to a lesser extent Red cedar, were not as fully represented as would have been desirable. Adjacent to each minor species plot, another assessment plot was established in the most productive available major species.

Most plots selected in this way occurred in pairs, i.e. one major and one minor species plot. However, where three or more relevant species were grouped together on a site, plots were established in all available species.

TABLE 29B
DISTRIBUTION OF PLOTS ASSESSED DURING SURVEY FIELDWORK

Meteorological Region	Corresponding Forestry Commission Conservancies	WH	RC	GF	NF	SP	CP	SS	NS	JL	HL	DF
0	Northern Scotland less Nairn and Moray Firth lowland*	7	4	8	7	—	—	8	6	—	—	5
1	East Scotland plus Nairn and Moray Firth lowland*	8	3	6	2	2	—	2	4	—	1	4
2	North-east England plus Lincolnshire	4	2	4	—	3	1	1	4	—	—	—
3	East England less Lincs., Beds., Northants., Oxon. and Bucks.	3	5	6	—	1	9	—	—	—	—	1
4	The Midlands, includes parts of N.W. England, S.W. England, E. England	4	6	6	1	1	2	2	1	1	—	5
5	South-east England, New and Wilts.	7	6	7	1	—	6	1	3	—	1	8
6a	West Scotland	9	2	3	6	—	—	9	7	1	—	—
6b	Western Counties of South Scotland	5	—	3	2	—	—	—	2	3	1	1
7a	North-west England counties, north from Cheshire	3	—	—	—	—	—	2	—	—	—	1
7b	North Wales less Cardigan and Rads.	6	7	7	3	—	1	7	3	—	—	8
8a	South Wales plus Cardigan and Rads.	8	6	9	4	—	1	6	9	5	1	4
8b	South-west England, west from Somerset and Dorset	8	7	8	—	—	—	6	—	1	—	9
	Total	72	48	67	26	7	20	44	39	11	4	46

* Data for Culloden, Black Isle and Novar Estate were also taken from Met. Reg. 0 and grouped with Met. Reg. 1 data.

A list of plots in numerical order is given in Table 30. Details of forest, compartment, species, planting year, yield class and soil type are given in the key to Figures 1 to 18 where the plot data is tabulated. See Chapter 3, pages 10 to 45.

Assessments in Plots in 1968 Survey

Within each stand, a small assessment plot was marked out, and the following assessed:

Years of planting
Number of stems per acre
Mean quarter-girth at breast-height } Based on trees in 0.10 acre plot
Stand basal area per acre
Girth of 40 largest stems per acre } Based on trees in 0.15 acre plot
Top height (ht of 40 trees of greatest girth, per acre

From this information (General) Yield Class and (Local) Production Class were determined. Figures were subsequently converted from Imperial to metric measures during the course of preparing this report.

The frequency and severity of the following defects were assessed in each stand:
Stem fluting
Stem crack
Other stem defects (e.g. crooked stems, double leaders, animal damage, butt sweep, etc.)
Pattern of windblow
Crown blast
Incidence of heart rot (on freshly cut stumps or thinnings still lying in the wood).

Cores taken from 20 trees in all stands over 35 years old and in all second generation conifer stands were sent to the Pathology Laboratory, Alice Holt, where they were incubated and examined for infection by *Fomes annosus*.

The previous crop or land use was noted, together with whether the site had been ploughed or drained and how often the present stand had been thinned.

Soils

One or more samples from each profile were analysed for pH per cent stones, per cent water, loss of weight on ignition, per cent sand, clay and silt, and total phosphate

APPENDIX I

TABLE 30
MINOR SPECIES SURVEY
LIST OF STANDS VISITED

Survey Stand No.	Conservancy	Forest	Compartment No.	Species	Used in Comparison with Plot Nos.

SECTION 1: METEOROLOGICAL REGION 5

Survey Stand No.	Conservancy	Forest	Compartment No.	Species	Used in Comparison with Plot Nos.
1	South East England	Alice Holt	63c	GF	2
2	,, ,, ,,	,, ,,	62e	CP	1
3	,, ,, ,,	,, ,,	38a	WH	4, 5, 6
4	,, ,, ,,	,, ,,	27a	CP	3, 5, 6
5	,, ,, ,,	,, ,,	29i	RC	3, 4, 6
6	,, ,, ,,	,, ,,	25b	DF	3, 4, 5
7	,, ,, ,,	,, ,,	53c	RC	8
8	,, ,, ,,	,, ,,	53e	DF	7
9	,, ,, ,,	Andover	8	WH	10
10	,, ,, ,,	,,	8	DF	9
11	,, ,, ,,	Abinger (Chiddingfold)	10	GF	12
12	,, ,, ,,	,, ,,	10b	NS	11
13	,, ,, ,,	Abinger (Chiddingfold, U. Sidney Wood)	67a	WH	14–17
14	,, ,, ,,	,, ,, ,,	68	RC	13, 15–17
15	,, ,, ,,	,, ,, ,,	68c	GF	13, 14, 16, 17
16	,, ,, ,,	,, ,, ,,	69a	WH	13–15, 17
17	,, ,, ,,	,, ,, ,,	69c	NS	13–16
18	South West England	Cranborne Chase	12	WH	19
19	,, ,, ,,	,, ,,	17	DF	18
20	South East England	Maresfield (Gravetye)	329(i)	GF	21
21	,, ,, ,,	,, ,,	335(a)	CP	20
22	,, ,, ,,	New Forest: Broomy	1/39a	WH	23
23	,, ,, ,,	,, Rhinefield	40c	CP	22
24	,, ,, ,,	,, ,,	21g	RC	25
25	,, ,, ,,	,, ,,	20c	DF	24
26	,, ,, ,,	Challock ((Orlestone)	7e	RC	27
27	,, ,, ,,	,, ,,	7c	DF	26
28	,, ,, ,,	Ringwood	29f	GF	29
29	,, ,, ,,	,,	28	CP	28
30	,, ,, ,,	Bedgebury (Vinehall)	23	GF	31
31	,, ,, ,,	,,	29	DF	30
250	,, ,, ,,	Bedgebury (Research Plot)	46	WH	251, 258, 259
251	,, ,, ,,	,, ,,	55	RC	250, 258, 259
252	,, ,, ,,	,, ,,	28	GF	253–257
253	,, ,, ,,	,, ,,	38	NF	252, 254–257
254	,, ,, ,,	,, ,,	15	NS	252/3, 255–257
255	,, ,, ,,	,, ,,	69	SS	252–254, 256/7
256	,, ,, ,,	,, ,,	98	DF	252–255, 257
257	,, ,, ,,	,, ,,	89	CP	252–256
258	,, ,, ,,	,, ,,	65	HL	250, 251, 259
259	,, ,, ,,	,, ,,	45	LC	250, 251, 258

SECTION 2: METEOROLOGICAL REGION 8b

Survey Stand No.	Conservancy	Forest	Compartment No.	Species	Used in Comparison with Plot Nos.
32	South West England	Bodmin	43/2/5d	WH	33, 34
33	,, ,, ,,	,,	43/2/5c	GF	32, 34
34	,, ,, ,,	,,	43/2/5a	JL	32, 33

TABLE 30 (cont.)

Survey Stand No.	Conservancy			Forest			Compartment No.	Species	Used in Comparison with Plot Nos.
35	South West England			Dartington Estate, Devon*			I	WH	36
36	,,	,,	,,	,,	,,	,,	I	DF	35
37	,,	,,	,,	,,	,,	,,	I	GF	38, 39
38	,,	,,	,,	,,	,,	,,	VI	RC	37, 39
39	,,	,,	,,	,,	,,	,,	IV	DF	37, 38
40	,,	,,	,,	Dartmoor			10a	WH	41
41	,,	,,	,,	,,			8b	SS	40
42	,,	,,	,,	Eggesford			25/27	RC	43
43	,,	,,	,,	,,			25/27	SS	42
44	,,	,,	,,	,,			2a	RC	45
45	,,	,,	,,	,,			1	DF	44
46	,,	,,	,,	Quantock (E. Quantoxhead), Somerset			—	WH	47
47	,,	,,	,,				—	DF	46
48	,,	,,	,,	Halwill			62/4/1d	WH	49, 50, 51
49	,,	,,	,,	,,			62/4/4e	SS	48, 50, 51
50	,,	,,	,,	,,			62/4/1c	LC	48, 49, 51
51	,,	,,	,,	,,			62/4/4d	GF	48, 49, 50
52	,,	,,	,,	Mendip			194a	GF	53
53	,,	,,	,,	,,			194c	DF	52
54	,,	,,	,,	Halwill (Okehampton)			64/4/14d	WH	55
55	,,	,,	,,	,,	,,		64/4/14b	DF	54
56	,,	,,	,,	,,	,,		64/4/9a	GF	57, 58
57	,,	,,	,,	,,	,,		64/4/8	LC	56, 58
58	,,	,,	,,	,,	,,		64/4/9c	SS	56, 57
59	,,	,,	,,	Dartmoor (Plym)			15	RC	60–63
60	,,	,,	,,	,,	,,		,,	GF	59, 61–63
61	,,	,,	,,	,,	,,		,,	WH	59, 60, 62, 63
62	,,	,,	,,	,,	,,		7	DF	59–61, 63
63	,,	,,	,,	,,	,,		8	GF	59–62
64	,,	,,	,,	Quantock			49/3/52	WH	65
65	,,	,,	,,	,,			,,	SS	64
66	,,	,,	,,	,,			49/3/23b	GF	67
67	,,	,,	,,	,,			49/3/22	SS	66
68	,,	,,	,,	,,			49/3/3d	RC	69
69	,,	,,	,,	,,			49/3/4	DF	68
70	,,	,,	,,	Tavistock Woodlands Ltd., Devon*			D2J	RC	71, 72
71	,,	,,	,,				DsF	DF	70, 72
72	,,	,,	,,	,,			D2M	RC	70, 71

SECTION 3: METEOROLOGICAL REGION 8a

73	South Wales		Brechfa (III)	472b	WH	74
74	,,	,,	,,	472a	JL	73
75	,,	,,	Brechfa (V)	833d	GF	76
76	,,	,,	,,	833c	DF	75
77	,,	,,	Caeo	46c	WH	78, 79
78	,,	,,	,,	46d	GF	77, 79
79	,,	,,	,,	46a	JL	77, 78
80	,,	,,	,,	33a	NS	81, 82
81	,,	,,	,,	33c	LC	80, 82
82	,,	,,	,,	33e	RC	80, 81

*Private estate.

APPENDIX I

TABLE 30 (cont.)

Survey Stand No.	Conservancy	Forest	Compart-ment No.	Species	Used in Comparison with Plot Nos.
83	South Wales	Crychan	25a	GF	84–86
84	,, ,,	,,	25b	SS	83, 85 86
85	,, ,,	,,	24a	JL	83, 84 86
86	,, ,,	,,	24f	RC	83–85
87	,, ,,	Wentwood (Draethen)	16b	RC	88
88	,, ,,	,, ,,	16a	CP	87
89	,, ,,	Tair Onen (Llantrisant)	11c	WH	90
90	,, ,,	,, ,, ,,	,,	DF	89
91	,, ,,	,, ,, ,,	11f	GF	92
92	,, ,,	,, ,, ,,	10f	SS	91
93	,, ,,	,, ,, ,,	18d	NF	94
94	,, ,,	,, ,, ,,	,,	NS	93
95	,, ,,	Mynydd Du	19d	WH	96
96	,, ,,	,, ,,	19c	NS	95
97	,, ,,	,, ,,	2c	RC	98
98	,, ,,	,, ,,	2a	SS	97
99	,, ,,	,, ,,	22a	DF	100–103
100	,, ,,	,, ,,	22d	GF	99, 101–103
101	,, ,,	,, ,,	22b	NF	99, 100, 102, 103
102	,, ,,	,, ,,	31a	NS	99–101, 103
103	,, ,,	,, ,,	22e	LC	99–102
104	,, ,,	Coed Abertawe	3a	JL	105
105	,, ,,	,, ,,	3b	GF	104
106	,, ,,	Talybont	7a	WH	107–111
107	,, ,,	,,	7f	RC	106–108–111
108	,, ,,	,,	3c	GF	106, 107, 109–111
109	,, ,,	,,	4c	GF	106–108, 110, 111
110	,, ,,	,,	3d	NS	106–109, 111
111	,, ,,	,,	8b	JL	106–110
112	,, ,,	,,	39a	WH	113
113	,, ,,	,,	38a	NS	112
114	,, ,,	,,	37b	NF	115
115	,, ,,	,,	37a	NS	114
116	,, ,,	Tintern	91d	WH	117
117	,, ,,	,,	93a	NS	116
118	,, ,,	Wentwood	1a	RC	119, 120
119	,, ,,	,,	1b	HL	118, 120
120	,, ,,	,,	1c	NS	118, 119

SECTION 4: METEOROLOGICAL REGION 7b

Survey Stand No.	Conservancy	Forest	Compart-ment No.	Species	Used in Comparison with Plot Nos.
121	North Wales	Coed-y-Brenin	P13j	WH	122
122	,, ,,	,, ,,	P13i	NS	121
123	,, ,,	,, ,,	S17–21	WH	124
124	,, ,,	,, ,,	,,	SS	123
125	,, ,,	,, ,,	R15d	GF	126, 127
126	,, ,,	,, ,,	R15c	RC	125, 127
127	,, ,,	,, ,,	R15a	DF	125, 126
128	,, ,,	,, ,,	P7f	GF	129
129	,, ,,	,, ,,	P8a	DF	128
130	,, ,,	Cynwyd	54	WH	131, 132
131	,, ,,	,,	—	SS	130, 132
132	,, ,,	,,	—	GF	130, 131

TABLE 30 (*cont.*)

Survey Stand No.	Conservancy		Forest	Compartment No.	Species	Used in Comparison with Plot Nos.
133	North Wales		Beddgelert (Deudraeth)	19d	WH	134, 135, 136/7
134	,,	,,	,, ,,	19c	DF	133, 135, 136/7
135	,,	,,	,, ,,	19j	RC	133/4, 136/7
136	,,	,,	,, ,,	20–21	SS	133/4, 135, 137
137	,,	,,	,, ,,	6a	NS	133/4, 135, 136
138	,,	,,	Dyfi Corris	41	GF	139, 140
139	,,	,,	,, ,,	41a	WH	138, 140
140	,,	,,	,, ,,	41b	DF	138, 139
141	,,	,,	,, ,,	95c	NF	142–144
142	,,	,,	,, ,,	95b	SS	141, 143/4
143	,,	,,	,, ,,	30b	RC	141/2, 144
144	,,	,,	,, ,,	30a	NS	141/143
147	,,	,,	Dyfi (Valley)	13a	NF	148
148	,,	,,	,, ,,	13b	SS	147
149	,,	,,	Gwydyr	164k	RC	150
150	,,	,,	,,	322a	SS	149
151	,,	,,	,,	199b	GF	152
152	,,	,,	,,	,,	DF	151
153	,,	,,	,,	206e	GF	154
154	,,	,,	,,	206b	DF	153
155	,,	,,	,,	320	GF	156–159
156	,,	,,	,,	,,	NF	155, 157–159
157	,,	,,	,,	,,	WH	155/6, 158/9
158	,,	,,	,,	321h	DF	155–157, 159
159	,,	,,	,,	321	CP	155–158
160	,,	,,	,,	331i	RC	161/ 162
161	,,	,,	,,	331j	LC	160, 161
162	,,	,,	,,	321	DF	160, 161
163	,,	,,	Radnor	72b	SS	164
164	,,	,,	,,	73c	NF	163
165	,,	,,	,,	56d	GF	166
166	,,	,,	,,	58b	SS	165
167	,,	,,	,,	66	WH	168, 204
168	,,	,,	,,	66d	DF	167, 204
204	,,	,,	,,	67b	SS	167, 168
169	,,	,,	Vyrnwy Woodlands,	5	RC	170
170	,,	,,	Montgomery	5	SS	169

SECTION 5: METEOROLOGICAL REGION 4

171	East England		Bagley Wood, Nr. Oxford*	15	RC	172–176
172			,, ,,	15	WH	172, 173–6
173			,, ,, ,,	15	JL	171/2, 174–6
174	,,	,,	,, ,, ,,	16	DF	171–3, 175/6
175	,,	,,	,, ,, ,,	16	NS	171–4, 176
176	,,	,,	,, ,, ,,	25e	GF	171–175
177	Forest of Dean (South-west England)		Abbotswood	438	LC	178
178	,,	,,	,,		DF	177
179	,,	,,	Blakeney	444g	GF	180
180	,,	,,	,,	444h	SS	179
181	,,	,,	Churchill	254c	GF	182–184
182	,,	,,	,,	,,	NF	181, 183/4
183	,,	,,	,,	,,	WH	181/2, 184
184	,,	,,	,,	254a	EL	181–183
185	,,	,,	,,	246g	GF	186–188
186	,,	,,	,,	246c	LC	185, 187, 188
187	,,	,,	,,	246d	SS	185/6, 188
188	,,	,,	,,	245c	CP	185–187

APPENDIX I

TABLE 30 (cont.)

Survey Stand No.	Conservancy	Forest	Compartment No.	Species	Used in Comparison with Plot Nos.
189	Forest of Dean (South-west England)	High Meadow	31b	RC	190
190	,, ,,	,, ,,	,,	LC	189
191	,, ,,	,, ,,	57	GF	192
192	,, ,,	,, ,,	,,	WH	191
193	,, ,,	Nagshead	79b	RC	194
194	,, ,,	,,	79e	DF	193
195	,, ,,	,,	80c	GF	196
196	,, ,,	,,	80b	DF	195
197	South-west England	Hereford (Dymock)	28	RC	198/9
198	,, ,,	,, ,,	21	WH	197/199
199	,, ,,	,, ,,	9b	DF	197/8
200	East England	Rockingham (Bedford Purlieus)	23b	RC	201
201	,, ,,	,, ,,	22d	CP	200
202	,, ,,	,, (Fineshade)	11	SP	203
203	,, ,,	,, ,,	22	RC	202

SECTION 6: METEOROLOGICAL REGION 3

Survey Stand No.	Conservancy	Forest	Compartment No.	Species	Used in Comparison with Plot Nos.
221	East England	Ampthill	2	RC	222/3
222	,, ,,	,,	2	GF	221/223
223	,, ,,	,,	5	CP	221/2
224	,, ,,	Thetford	16	CP	225
225	,, ,,	,,	55	GF	224
226	,, ,,	,,	157	GF	227–229
227	,, ,,	,,	157	RC	226, 228/9
228	,, ,,	,,	157	WH	226/7, 229
229	,, ,,	,,	158	CP	226–228
230	,, ,,	Wensum (Swanton)	36	GF	231/2
231	,, ,,	,, ,,	31	DF	230, 232
232	,, ,,	,, ,,	33	CP	230, 231
233	,, ,,	,, (Holt)	26	GF	234
234	,, ,,	,, ,,	26	CP	233
235	,, ,,	North Lindsey (Willingham)	17c	WH	236–237
236	,, ,,	,, ,, ,,	17a	SP	235, 237
237	,, ,,	,, ,, ,,	21	CP	235/6
238	,, ,,	Kesteven	78c	RC	239
239	,, ,,	,,	78b	CP	238
240	,, ,,	,,	39d	RC	241
241	,, ,,	,,	39d	GF	240
243	,, ,,	,,	28	RC	244
244	,, ,,	,,	28	CP	243
245	,, ,,	,,	28	WH	246
246	,, ,,	,,	28	CP	245

Note. Plots 250–259 were located at Bedgebury Pinetum and are shown, following plot 31, along with other plots in South-east England.

SECTION 7: METEOROLOGICAL REGION 0

Survey Stand No.	Conservancy	Forest	Compartment No.	Species	Used in Comparison with Plot Nos.
501	North Scotland	Leanachan (Clunes)	30h	WH	502
502	,, ,,	,, ,,	30g	NS	501

*Private estate.

TABLE 30 (*cont.*)

Survey Stand No.	Conservancy	Forest	Compartment No.	Species	Used in Comparison with Plot Nos.
503	North Scotland	Corrour (Fersit)	8a	RC	Nil
504	,, ,,	Ratagan (Eileanreach)	37i	GF	505
505	,, ,,	,, ,,	37j	SS	504
506	,, ,,	Leanachan (Clunes:	59g	WH	507
507	,, ,,	Glen Loy)	59a	NS	506
508	,, ,,	Glen Urquhart	17a	WH	509
509	,, ,,	,, ,,	17b	DF	508
510	,, ,,	,, ,,	36d	RC	511
511	,, ,,	,, ,,	33a/b	DF	510
512	,, ,,	,, ,,	25f	GF	513
513	,, ,,	,, ,,	25a/g	SS	512
514	,, ,,	,, ,,	69b/c	NF	515
515	,, ,,	,, ,,	65f/g	DF	514
516	,, ,,	,, ,,	5f	NF	517
517	,, ,,	,, ,,	5h	DF	516
518	,, ,,	,, ,,	5f	NF	519
519	,, ,,	,, ,,	5b	SS	518
520	,, ,,	Inchnacardoch	36a	GF	521, 522
521	,, ,,	,,	36e	NS	520, 522
522	,, ,,	,,	36f	NF	520–521
523	,, ,,	Ratagan (Inverinate)	16f	WH	524, 525
524	,, ,,	,, ,,	15d	SS	523, 525
525	,, ,,	,, ,,	16e	RC	523, 524
526	,, ,,	Torrachilty (Lael)	61e	WH	527
527	,, ,,	,, ,,	61d	SS	526
528	,, ,,	Inchnacardoch (Portclair)	14a	WH	529–532
529	,, ,,	,, ,,	14b	DF	528, 530–532
530	,, ,,	,, ,,	14e	RC	528/9, 531/2
531	,, ,,	,, ,,	14g	GF	528–530, 532
532	,, ,,	,, ,,	14f	NF	528–531
533	,, ,,	,, ,,	19d	GF	534
534	,, ,,	,, ,,	24b	SS	533
535	,, ,,	,, ,,	71d	GF	536
536	,, ,,	,, ,,	71b	SS	535
537	,, ,,	Ratagan	40bk	GF	538, 539
538	,, ,,	,,	40bk	NS	537, 539
539	,, ,,	,,	40bj	NF	537, 538
540	,, ,,	,,	40ac	NF	541
541	,, ,,	,,	43g	SS	540
542	,, ,,	Leanachan	60	WH	543
543	,, ,,	,,	60	NS	542
544	,, ,,	Inchnacardoch	49d	GF	545
545	,, ,,	,,	49e	NS	544

SECTION 8: METEOROLOGICAL REGION 1

Survey Stand No.	Conservancy	Forest	Compartment No.	Species	Used in Comparison with Plot Nos.
601	East Scotland	Alltcailleach	22	GF	602
602	,, ,,	,,	22	DF	601
603	,, ,,	,,	18	NF	604
604	,, ,,	,,	21	SP	603
605	,, ,,	Bennachie	16	WH	Nil

APPENDIX I

TABLE 30 (cont.)

Survey Stand No.	Conservancy	Forest	Compartment No.	Species	Used in Comparison with Plot Nos.
606	East Scotland	Strathardle (Blackcraig)	6	WH	607
607	,, ,,	,, ,,	6	NS	606
608	North Scotland	Black Isle	18	WH	609
609	,, ,,	,, ,,	18	DF	608
614	East Scotland	Dunkeld (Craigvinean)	83	GF	615
615	,, ,,	,, ,,	83	NS	614
616	North Scotland	Culloden	25c	WH	617
617	,, ,,	,,	25b	DF	616
618	,, ,,	,,	201	RC	619
619	,, ,,	,,	201	DF	618
620	East Scotland	Darnaway Estate Moray*	6	RC	621
621	,, ,,	,, ,, ,,	6	SP	620
622	,, ,,	Drummond Hill	46	WH	623
623	,, ,,	,, ,,	47	HL	622
624	,, ,,	,, ,,	97	GF	625
625	,, ,,	,, ,,	57	NS	624
630	,, ,,	Mearns (Drumtochty)	66b	GF	631, 632
631	,, ,,	,, ,,	66d	NS	630, 632
632	,, ,,	,, ,,	66c	NF	630, 631
635	,, ,,	Montreathmont (Inglismaldie)	19	WH	636, 637
636	,, ,,	,, ,,	18	SS	635, 637
637	,, ,,	,, ,,	19	GF	635, 636
638	,, ,,	Laigh of Moray (Monaughty)	80	WH	639–641
639	,, ,,	,, ,, ,,	81	SS	638, 640/1
640	,, ,,	,, ,, ,,	80	RC	638/9, 641
641	,, ,,	,, ,, ,,	72	GF	638–640
642	North Scotland	Novar Estate, Ross-shire*	81/85	WH	Nil

SECTION 9: METEOROLOGICAL REGION 6a

Survey Stand No.	Conservancy	Forest	Compartment No.	Species	Used in Comparison with Plot Nos.
701	West Scotland	Achaglachgach	48	WH	702
702	,, ,,	,,	,,	SS	701
703	,, ,,	,,	76	WH	704
704	,, ,,	,,	,,	NS	703
705	,, ,,	,,	43	WH	706
706	,, ,,	,,	,,	NS	705
707	,, ,,	,,	41/44	GF	708
708	,, ,,	,,	,,	NS	707
709	,, ,,	Benmore	133	NF	710
710	,, ,,	,,	,,	SS	708
711	,, ,,	Eredine	86	WH	712
712	,, ,,	,,	,,	SS	711
713	,, ,,	,,	80	GF	714
714	,, ,,	,,	,,	NS	713
718	,, ,,	Glenbranter	48	NF	719
719	,, ,,	,,	,,	SS	718
720	,, ,,	Glenfinart	20	GF	721, 722
721	,, ,,	,,	16	SS	720, 722
722	,, ,,	,,	21	WH	720, 721

*Private estate

TABLE 30 (cont.)

Survey Stand No.	Conservancy	Forest	Compartment No.	Species	Used in Comparison with Plot Nos.
723	West Scotland	Inverliever	75	RC	724
724	,, ,,	,,	75	JL	723
725	,, ,,	Knapdale	13c	WH	726, 727
726	,, ,,	,,	13a	SS	725, 727
727	,, ,,	,,	13d	NF	725, 726
732	,, ,,	Loch Ard	83	WH	733
733	,, ,,	,, ,,	,,	NS	732
734	,, ,,	,, ,,	79	RC	Nil
736	,, ,,	Glenbranter (Loch Eck)	52	WH	737
737	,, ,,	,, ,, ,,	54	NS	736
739	,, ,,	,, ,, ,,	96	NF	740
740	,, ,,	,, ,, ,,	,,	SS	739
741	,, ,,	,, ,, ,,	7	NF	742
742	,, ,,	,, ,, ,,	,,	SS	741
746	,, ,,	Fearnoch	27	NF	747
747	,, ,,	,,	,,	SS	746
748	,, ,,	Inverliever (Inverinan)		WH	749
749	,, ,,	,, ,,		NS	748
801	South Scotland	Solway (Dalbeattie)	415	WH	802
802	,, ,,	,, ,,	,,	DF	801
803	,, ,,	,, ,,	77	WH	804
804	,, ,,	,, ,,	,,	JL	803
805	,, ,,	Fleet	22	GF	806
806	,, ,,	,,	22	JL	805
813	,, ,,	Kirroughtree	35	WH	814
814	,, ,,	,,	19	EL	813
815	,, ,,	,,	39	GF	816
816	,, ,,	,,	,,	JL	815
817	,, ,,	Fleet (Laurieston)	4	WH	818
818	,, ,,	,, ,,	4	HL	817
820	,, ,,	Solway (Mabie)	24	GF	821
821	,, ,,	,, ,,	,,	NS	820
822	,, ,,	Queensberry Estate*	24	WH	823, 824
823	,, ,,	,, ,,	,,	NS	822, 824
824	,, ,,	,, ,,	,,	NF	822, 823
825	,, ,,	,, ,,	,,	NF	Nil

SECTION 10: METEOROLOGICAL REGION 2

901	North East England	Allerston District (Allerston)	SS25f	WH	902, 903
902	,, ,, ,,	,, ,,	—	NS	901, 903
903	,, ,, ,,	,, ,,	SS25e	GF	901, 902
904	,, ,, ,,	Dechant Estate*	Naboth 4	RC	905
905	,, ,, ,,	,, ,,	,,	SP	904
906	,, ,, ,,	Ford Estate,* Northumberland	(Fenwick Wood)	RC	907
907	,, ,, ,,			CP	906
908	,, ,, ,,	Hambleton (Ampleforth)	196	WH	909
909	,, ,, ,,	,, ,,		SP	908
910	,, ,, ,,	,, ,,		GF	911
911	,, ,, ,,	,, ,,	9	NS	910

*Private Estate

APPENDIX I

TABLE 30 (*cont.*)

Survey Stand No.	Conservancy	Forest	Compartment No.	Species	Used in Comparison with Plot Nos.
912	North East England	Hamsterley	1a	WH	913
913	,, ,, ,,	,,	,,	SP	912
914	,, ,, ,,	Kielder District (Kielder)	1348	WH	915
915	,, , ,,	,, ,, ,,	1350	SS	914
916	,, ,, ,,	,, ,, ,,	1330	GF	917
917	,, ,, ,,	,, ,, ,,	,,	NS	916
918	,, ,, ,,	,, ,, ,,	1144	GF	919
919	,, ,, ,,	,, ,, ,,	,,	NS	918
920	,, ,, ,,	Kyloe Estate, Northumberland*	4a	RC	Nil

SECTION 11: METEOROLOGICAL REGION 7a

953	North West England	Dunnerdale	6	WH	954
954	,, ,, ,,	,,	6	DF	953
955	,, ,, ,,	Grizedale	119	WH	956
956	,, ,, ,,	,,	118	SS	955
959	,, ,, ,,	Thornthwaite	343a	WH	960
960	,, ,, ,,	,,	343b	SS	959

*Private estate

APPENDIX II

REPORT ON SURVEY INTO TIMBER LOSSES IN WESTERN HEMLOCK AND GRAND FIR AS A RESULT OF INFECTION BY FOMES ANNOSUS

By J. E. PRATT,
Pathology Section, Research Division, Forestry Commission

Summary
A preliminary study was made into the relative susceptibility of Western hemlock and Grand fir to staining and decay caused by *Fomes annosus*. Although not complete, the following points emerged:
  (i) Grand fir is relatively resistant and losses were negligible.
  (ii) Western hemlock is susceptible; on heavily infested sites, losses in timber volume in butts of infected trees were about 30 per cent while in the crop as a whole, up to 20 per cent of the total volume can be described as cull.
  (iii) Severe losses occurred on sites with a previous history of conifers. An important factor influencing loss, at least as revealed in studies of Western hemlock, seems to be the quantity and distribution of infection sources throughout the previous crop, and these can vary considerably from site to site. The relationship of these infection sources to future replacement planting programmes for Western hemlock is briefly discussed.

Introduction
An assessment of the growth of some minor species in British forests was made in 1968. This survey included preliminary investigations into the comparative susceptibility of minor and major species to infection by *Fomes annosus*. The investigation was based on small sampling units, and comparisons between species could not be relied on. As a consequence, a more detailed study into the actual losses in two species, Western hemlock and Grand fir was undertaken.

Objects
  (i) To compare the susceptibility of Grand fir and Western hemlock to *Fomes* butt rot.
  (ii) To measure volume loss (or cull) in infected stems and estimate the maximum loss to be expected in a variety of age classes.

Method
  (i) *Site selection*. Sites studied were all second rotation conifer sites with a history of heavy infection either in the previous crop or in the present crop. Most of these sites were selected with the aid of the *Fomes* survey notes of Gladman and Low, 1955–1960. A few came from the minor species survey. An indication of the level of attack by *Fomes* at a given site was obtained by an examination of stumps of the previous crop and of thinning stumps of the current crop. There are limitations to the use of this method of assessment as the identification of *Fomes* infection in stumps under forest conditions is not straight-forward.
  (ii) *Tree selection*. Pressler borings were taken at about 8 in. above ground from 50 trees selected at

TABLE 31
LIST OF SITES

Species	Site No.	Forest	Compartment	Planting Year	Age	Area (hectares)	Number of Thinnings	Av. size cu m (approx)
WH	1	Dean	434/5	51	18	1·2	Nil	0·11
GF	1	Dean	434/5	51	18	1·2	Nil	0·11
WH	2	Drummond Hill	48	27	42	1·6	3	0·32
WH	3	Monaughty(¹)	80	26	43	1·6	4	0·36
GF	3	Monaughty(¹)	72	26	43	2·0	7	0·58
WH	4	Monaughty(¹)	63	33	36	2·0	2*	0·29
GF	4	Monaughty(¹)	63	33	36	2·0	1*	0·36
WH	5	Drumtochty(²)	61(b)	34	35	3·2	3	0·25
GF	5	Drumtochty(²)	61(b)	34	35	0·4		0·29*
WH	6	Kessock(³)	106/8	28	41	4·0	6	0·36
GF	7	Bin(⁴)	36	34	35	2·4	3*	0·50*

Notes:
*estimated figures
(¹)Now part of Laigh of Moray Forest
(²)Now part of Mearns Forest
(³)Now part of Black Isle Forest
(⁴)Now part of Huntly Forest

random throughout each site. *Fomes* was identified by incubating or culturing the cores. A proportion of the affected trees were felled. The limits of stain and decay in each tree were determined by progressive cross-cutting from the butt, and the total volume affected was measured. Stems with external damage or with signs of *Armillaria mellea* attack were specifically excluded from the study.

Results
(i) **Sites visited**

Table 31 lists sites where incidence and, in most cases, loss were assessed. The susceptibility of the two species can be compared where the site numbers are the same. Other sites visited had insufficient infection to justy more thorough assessments.

Most of the sites fall into the 35–43 year old age group. Older crops with enough trees for a full-scale study were not found.

(ii) **Fomes Incidence in Present Crop**

In all cases except at Kessock (site 6), Black Isle Forest, incidence within the crop was determined by taking Pressler borings. The accuracy of this method has not been fully tested under all circumstances and can only give an estimate of the number of infected trees in a stand. However, it is the only non-destructive method of sampling available, and so far as comparisons between different species are feasible, it gives a guide. This assessment was used primarily to locate trees with internal decay. The results from assessments of boring are given in Table 32.

TABLE 32
FOMES INCIDENCE PER CENT

Site No.	1	2	3	4	5	6	7
Western hemlock	14	46	50	18	8	96*	NC
Grand fir	0	NC	8	0	2	NC	0

*Records from 50 trees felled during thinning operations.
NC – No trees available for comparison.

The table above shows percentage incidence figures for the sites studied. At sites 1, 3, 4 and 5 comparisons between adjacent or mixed crops of hemlock and Grand fir can be made. No direct comparison could be made at sites 2, 6 and 7. Western hemlock has shown consistently higher incidence than Grand fir. *Fomes* was found to be present in hemlock of various ages, in thinned stands and even in one unthinned stand. The high incidence at Kessock (site 6) is probably the result of early root contact with the large numbers of *Fomes* infected birch stumps, which were a legacy from the 17 years between successive conifer crops. The incidence at Monaughty (Laigh of Moray Forest) (site 4) is low and is confined to one small area of the compartment. However, decay of infected trees here is nearly as bad as at Kessock, Black Isle Forest, which suggests that infection took place early in the crop's life, but has been restricted to one area. The other sites had a previous history of either European larch or Scots pine, an important factor in determining the level of infection which develops in the current crop.

(iii) **Timber Loss**

a. *Decay in Western hemlock Sample Trees*

Table 33 shows details of loss by *decay* in various hemlock stands. Loss by decay is defined as the total volume of a length of timber in which there is a column of decay; it is expressed in cubic metres and also as a percentage of the total volume of the tree to 7.5 cm top diameter over bark. It is not an actual measure of the volume of the column of infection. The vertical extent of decay is expressed in metres and as a percentage of the tree height to 7.5 cm diameter.

The 'Approx. % Loss in Crop' figures in the last column are based on the average cull %, the average volume per tree in the standing crop, and the incidence figures derived from the initial survey by boring. It is, therefore, by no means precise.

TABLE 33
MAXIMUM AND AVERAGE LOSSES (DECAY ONLY)

Species	Age Years	Site No.	No of decayed trees measured	Loss by decay Greatest %	Loss by decay Greatest cu m	Loss by decay Average %	Loss by decay Average cu m	Vertical Extent Highest %	Vertical Extent Highest m	Vertical Extent Average %	Vertical Extent Average m	Approx. % loss in crop	Remarks
WH	18	1	1	negligible				0.15		0.15		negligible	v. early incipient decay
WH	42	2	7	27	0.09	14	0.04	14	1.8	7	0.9	3	
WH	43	3	7	25	0.08	14	0.06	12	1.8	8	1.1	5	
WH	36	4	4	30	0.09	26	0.08	20	2.4	14	1.7	2	
WH	28	6	14	58	0.07	25	0.10	25	3.5	16	2.3	10	

TABLE 34
MAXIMUM AND AVERAGE LOSSES
DECAY + STAIN

Spp	Site No.	No. of affected trees measured	Cull				Vertical Extent				Approx. % loss in crop		
			Greatest		Average		Highest		Average		Decay	Stain	Decay and Stain
			%	m³	%	m³	%	m	%	m			
WH	1	5	20	0·02	14	0·02	15	1·2	11	0·8	—	2	2
WH	2	15	39	0·16	20	0·05	21	3·0	11	1·4	3	6	9
WH	3	11	35	0·11	21	0·06	18	2·7	12	1·6	5	5	10
WH	4	7	37	0·10	29	0·08	29	3·6	17	1·8	2	2	4
WH	6	14									10	5–10* est.	15–20* est.

Decay includes both advanced and incipient decays, which were defined as follows:

(i) Incipient decay: Decay where the structure of the wood has been visibly altered due to fungal attack but where the annual rings are still distinct.

(ii) Advanced decay: Decay in which destruction of the wood has progressed to the point of partial or complete disintegration and the annual rings are no longer discernible.

The decay in the single hemlock tree measured on site 1 (Forest of Dean) was so small that no measurements were taken of its volume. This stand was only 18 years old.

b. *Decay in Grand fir Sample Trees*
No decay was found in Grand fir.

c. *Decay + Stain in Western Hemlock Sample Trees*

Table 34 shows measurements of stained timber added to those for decay to give a figure for combined loss. Measurements of stain were taken from trees with stain accompanying decay and on trees with stain only. For this reason the number of measured trees recorded in Table 34 is greater than that in Table 33. As well as loss per tree, an estimate is given in the table of the percentage loss in the crop. Sites 2, 3 and 4 are all in the 35–43 year age group, and it appears from Table 34 that the average losses do not vary markedly from one another. Data for hemlock from site 6 are included in the table although there are no measurements of stain for this site. It is estimated that the approximate percentage loss in crop for this stand would be between 15 and 20 per cent if stain were included. The crop on site 1 is 18 years old; although the approximate per cent loss in this stand is smaller than at the other sites, the average percentage loss per tree is not greatly different.

The mean volume of decay in decayed Western hemlock in the 35–43 year age group was 21.5 per cent, extending up the trunk to a height, on average, of 1.6 m. The mean amount of stain in trees not exhibiting decay was 9 per cent by volume with an average extent of 0.8 m. Comparable figure for young and older crops are not available.

d. *Decay + Stain in Grand fir Sample Trees*

Only one Grand fir felled in the whole study was infected with *Fomes*. This tree, on site 3 (Monaughty Cpt. 72, aged 43) (Laigh of Moray Forest) was only stained. The stain extended 1.8 m up the stem (10 per cent of timber height) and the volume affected was 0.08 cu m (20 per cent of total volume). This represented a loss of 0.2 per cent in the crop. Ten other Grand fir were felled during the survey to confirm the indication from the preliminary borings, that there was no infection present.

Discussion

Low and Gladman (Forestry Commission Leaflet No. 5) found that the most serious attacks of *Fomes* butt rot occurred on sites with a previous history of conifers. The crops studied in this report were planted on sites mainly with a previous history of Scots pine or European larch both of which were very probably grown on a long rotation. The final crop of trees in such plantations were therefore likely to be larger and at a wider spacing than would currently be found. Infection sources could therefore have been more widely separated than would be accepted in present day silviculture. However, the closer spacing of current final crops could result in more infection sources and hence more infected trees in future crops.

This survey of losses in hemlock and Grand fir is to be extended. It has already, however, done much to support the observations that Grand fir is relatively resistant to decay by *Fomes* whereas Western hemlock is susceptible to decay.

APPENDIX III
TREE STABILITY

By J. E. EVERARD
Silviculture (South) Section, Research Division, Forestry Commission

Introduction

In order to obtain some indication of the relative stability of minor species, tree pulling studies were carried out at six forests using the technique described in Forestry Commission Bulletin 40 (Fraser and Gardiner, 1967).

Table 35 lists the forests, the soil type and the species tested in each forest. Altogether twenty tests were made, five on Western hemlock, four on Grand fir, three each on Sitka spruce and Norway spruce, two on Douglas fir and one each on Noble fir, Western red cedar and Corsican pine.

TABLE 35
SPECIES STUDIED BY FOREST AND SOIL TYPE

Forest	Soil Type	Species
Mynydd Du	Brown earth	WH, GF, NF, NS, DF
Eredine	Shallow brown earth	WH, SS
Alice Holt	Surface water gley	WH, RC, GF, DF, CP
Halwill	Surface water gley	WH, GF, SS, NS
Thornthwaite	Surface water gley	WH, SS
Kielder	Peaty gley	GF, NS

It had been hoped to select sites so that comparison could always be made using data from local Sitka spruce, but this proved to be possible in only three of the six forests.

At each site, eight trees selected from dominant and co-dominants, were pulled over following the technique detailed in Chapter 1 of Bulletin 40.

Basis of Comparisons

Trees are overthrown as a result of the force of the wind on the crown generating such turning moments about the root of the tree that the tree is uprooted (or alternatively, that the stem snaps). The techniques used in this study, based on those described in Bulletin 40, assume that the forces applied by wind at a given speed to trees of a similar size are the same regardless of species. This assumption is discussed at the end of this paper.

Linear regressions of turning moment on stem weight have been calculated by the method of least squares and have been analysed in three ways.

The first analysis has been of the correlation between the linear regressions and the figures on which the regression is based. Where significant correlations have been obtained, the lines can be viewed with reasonable confidence. Where the correlation is not significant, the regression lines are unreliable. Thirteen out of 20 regressions were significantly correlated with their data. The results of this analysis are given in Table 36.

TABLE 36
CORRELATIONS BETWEEN DATA OBTAINED AND LINEAR REGRESSIONS BEST FITTED TO DATA

Forest	Regressions Significantly Correlated to Data 1% probability	Regressions Significantly Correlated to Data 5% probability	Regressions not Significantly Correlated
Mynydd Du	GF, DF	NS, WH	NF
Inverliever (Eredine)	SS	WH	—
Alice Holt	RC	GF, DF	WH, CP
Halwill	GF	—	NS, SS, WH
Thornthwaite	SS	WH	—
Kielder (District)	—	NS	GF

In the second analysis, the regressions obtained for different species at any one site were compared with each other. The results, given in Table 37, show that very few pairs of regression differed significantly.

TABLE 37
RESULTS OF COMPARISONS OF REGRESSIONS OF DIFFERENT SPECIES ON ANY GIVEN SITE.

(In this Table only those Results where one Species Differed Significantly from Another are Given)

Forest	Differences significant at 1%	Differences significant at 5%
Mynydd Du	—	WH compared with GF
		DF compared with GF
Alice Holt	—	DF compared with GF

Thirdly, the regressions obtained from each site were compared with the 'standard' regression for Sitka spruce from Bulletin 40, figure 5, making the assumption that on the soil of given site, the values of the Sitka spruce regressions in Bulletin 40 for that soil type would apply. The results of such comparisons are given in Table 38. It may be noted that the Sitka spruce studies at Thornthwaite and Halwill agreed reasonably with the 'standard' Bulletin 40 values for Sitka spruce, but that the regression at Eredine was much steeper than the 'standard'.

Regression of Turning Moment on Stem Weight

The regression lines of turning moment on stem weight for all Grand fir and Western hemlock plots, and also for all species at Alice Holt and at Mynydd Du are illustrated in figures 22 to 25.

Two sorts of difference can be seen from these figures. One where the slope of the regression differs and the other

TABLE 38
RESULTS OF COMPARISONS OF REGRESSIONS OBTAINED FROM THE SITES LISTED WITH THE REGRESSION LINES FOR SITKA SPRUCE GIVEN IN FIGURE 5, BULLETIN 40

Forest	Differences significant at 1%	Differences significant at 5%	Differences not significant
Mynydd Du Brown Earth	DF, GF	NS, WH	NF
Eredine Shallow brown earth	—	WH	SS
Alice Holt Surface water gley/ shallow brown earth	RC	GF, DF	WH, CP
Halwill Surface water gley	—	—	GF, WH, SS, NS
Thornthwaite	WH	—	SS
Kielder (District)	—	—	GF, NS

where two lines are more or less parallel but are widely separated.

The line for Grand fir, figure 22, being steeper than others, implies that the bigger the trees of this species on this site at Mynydd Du, the less susceptible to blow-down they are, relative to other species.

The line for Douglas fir at Mynydd Du has more or less the same slope as that for hemlock (Fig. 23) at the same forest but lies below it, implying that Douglas fir here is less resistant than hemlock to overturning, over the range of sizes of tree compared.

Comparisons of Regressions

Grand fir

On the brown earth at Mynydd Du the regressions between Grand fir and Western hemlock differ significantly in slope. Over the range of sizes of tree encountered, the Grand fir and hemlock regression lines cross, implying that hemlock is more stable when small but that Grand fir is more stable when crops are taller, the cross-over occurring at a stem weight corresponding to a height of 65 feet.

The regression line for Grand fir at Mynydd Du was significantly steeper than that for the 'standard' Sitka spruce on deep brown earths in Bulletin 40, and lies below it, implying that Grand fir is markedly less resistant to overthrow than 'standard' Sitka spruce when the crops are smaller but that by the time they are 90–100 feet tall there is little between them.

Of the two comparisons of Douglas fir and Grand fir, that of trees at Mynydd Du on a deep brown earth indicates that Douglas fir test trees were overturned by a lesser force than was required for Grand fir, though differences in the slope of the ground of the two sites may have exaggerated the difference. At Alice Holt, on soils intermediate between a surface water gley and a shallow brown earth, no sound conclusions can be reached because the Douglas fir test trees were so much smaller than their Grand fir counterparts.

Only at Halwill (surface water gley) were Sitka spruce and Grand fir studied on the same site. There were no indications of any difference between them.

Grand fir and Norway spruce were studied on similar soils at Mynydd Du, Halwill and at Kielder. Although of similar age, the Grand fir stems were larger than their counterparts in the Norway spruce plots. Comparisons, as far as they could be realistically made, indicated no consistent differences between the species.

Noble Fir

Only one study was carried out on this species. This, at Mynydd Du, gave inconsistent results and the regression was not significant. However, the values obtained from the study lie below most of those for other species obtained at Mynydd Du, and while the microtopography of the site may be responsible for some of this (the site was to some extent terraced), the data as they stand give no grounds for any claims that Noble fir has any greater resistance to overthrow than other species.

Western Hemlock

Hemlock was studied at five of the six forests where trees were pulled down. On four of the five sites, the hemlock regression is less steep than for almost all other species present. At the fifth site, Thornthwaite, the trees were so small that they cannot be validly included in such comparisons. These results suggest that with increasing size, individual hemlock trees can be overturned by relatively smaller applied wind forces than other species.

Figure 24 shows that the regressions for the three sites on shallow brown earths or suface water gleys lie remarkably close together, even though two of the three regression lines are not statistically significant. On the deep brown earth at Mynydd Du, the slope of the hemlock regression is significantly less than that of the Sitka spruce 'standard' in Bulletin 40 and also less than the Grand fir near the hemlock. Apart from this, none of the hemlock regressions differs significantly from those of other species.

Major Species

All the results for Sitka spruce, Norway spruce and Douglas fir have been discussed in preceding paragraphs in the comparison of minor species. One study was set up in a Corsican pine stand at Alice Holt Forest but was only partially successful in that five out of the eight trees pulled over snapped instead of being uprooted. The regression shown in Fig. 25 for Corsican pine is based on results from three trees, which represent the least firmly rooted of those in the plot examined.

Discussion

This has been the first study comparing the resistance of a range of species to overturning, using the tree pulling technique. The most important results are possibly

APPENDIX III

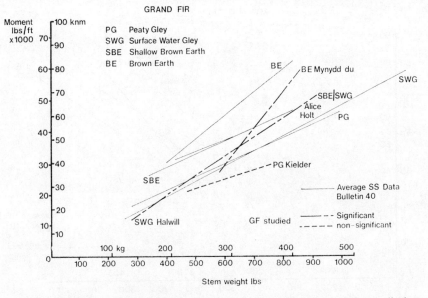

Figure 22. Regression lines of turning moment on stem weight for Grand fir, at all plots studied.

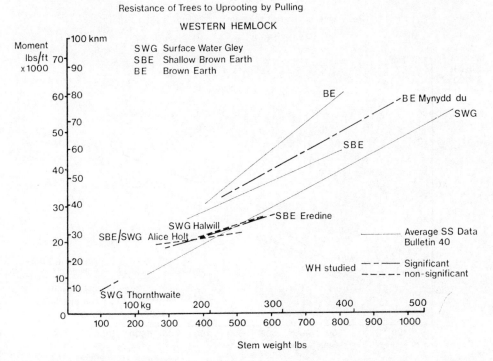

Figure 23. Regression lines of turning moment on stem weight for Western hemlock, at all plots studied.

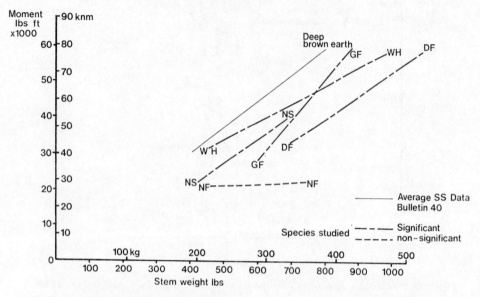

Figure 24. Regression lines of turning moment on stem weight for all species studied at Mynydd Du Forest.

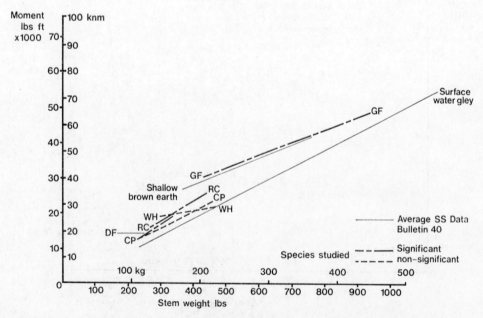

Figure 25. Regression lines of turning moment on stem weight for all species studied at Alice Holt Forest

APPENDIX III

negative – that no clear evidence was obtained that any species was markedly more resistant to overthrow than another.

Possibly the most consistent and well-founded observation is that hemlock regression lines were less steep than for almost all other species. Thus, in comparison with Grand fir it might be concluded that the relative stability of the two species depended on size. (The steeper slopes of the regressions for Grand fir imply it does become relatively more stable with age; the hemlock regressions on the other hand are markedly less steep, suggesting that as the trees get larger, they are relatively less resistant to overthrow.) However, it is pointed out in Section 2.23 that there is a greater range in heights in Grand fir stands than most and that Western hemlock stands are by contrast relatively uniform in height. This could be taken as implying that individually the dominants and co-dominantes in Grand fir stands are relatively less sheltered and therefore more subject to wind pressures than their counterparts in hemlock stands. The extent to which these opposing influences cancel each other out has yet to be determined.

Wind tunnel studies, albeit limited, indicate that the observable pliability of Western hemlock crowns permit more 'streamlining' of the crown in response to a given wind pressure than other species, and result in lower drag coefficient than for other species. (Mayhead, 1973.)

The only conclusions that can be reached are that for Grand fir and hemlock, small differences in relative resistance to overthrow may exist, but that their importance is not clear, and that soil type has a greater effect on stability than species.

As regards Noble fir and Red cedar, the data is entirely inadequate to support any conclusions.

REFERENCE

Fraser, A. I. and Gardiner J. B. H. (1967) Rooting and Stability in Sitka Spruce. *Bull. For. Comm., Lond.* 40.